Biotech Animals in Research

This book explores central aspects of genetic modification of animals for scientific purposes in the context of technological possibilities, regulatory issues in different regions, animal welfare implications and wider ethical issues exemplified through current theories and frameworks.

This discussion of lab animals produced through modern biotechnologies becomes increasingly pressing as CRISPR-Cas9 technology advances rapidly, challenging legal and ethical frameworks all over the world. Such animals are now affordable and readily available to almost every branch of scientific research. This not only raises enormous potential for creating 'tailored' models for human diseases but also rubs up against the traditional guiding principles (the 3Rs) for the humane use of animals for scientific experiments and raises wider ethical issues around death, integrity and naturalness. In this book, expert authors from diverse backgrounds in laboratory animal care, animal research, technology and animal rights explore a range of topics, from the science behind biotech research animals and the regulation of their use, to utilitarian, animal rights, virtue ethics and ethics of care, and critical animal studiers' perspectives on the use of these technologies.

Whatever your background or role in animal research, this book will challenge and stimulate deeper consideration of the benefits, disadvantages and ethical consequences of the use of biotechnology in the animal laboratory.

Biotech Animals in Research

Ethical and Regulatory Aspects

Mickey Gjerris, Anna Kornum,
Helena Röcklinsberg and Dorte Bratbo Sørensen

CRC Press
Taylor & Francis Group
Boca Raton London New York

CRC Press is an imprint of the
Taylor & Francis Group, an **informa** business

Cover image credit: Anna Kornum

First edition published 2024
by CRC Press
6000 Broken Sound Parkway NW, Suite 300, Boca Raton, FL 33487-2742

and by CRC Press
4 Park Square, Milton Park, Abingdon, Oxon, OX14 4RN

CRC Press is an imprint of Taylor & Francis Group, LLC

ISBN: 9781138369214 (hbk)
ISBN: 9781138369191 (pbk)
ISBN: 9780429428845 (ebk)

DOI: 10.1201/9780429428845

Typeset in Minion
by codeMantra

Contents

Authors

Mickey Gjerris (MTh, PhD in bioethics) is an associate professor at the Department of Food and Resource Economics at the University of Copenhagen, where he researches and teaches animal ethics, environmental ethics, bioethics, medical ethics and climate ethics. He teaches courses **for** veterinarians, animal researchers and animal caretakers, both at universities and at private companies. He is a former member at the Danish Council on Ethics.

Anna Kornum (MSc biology and philosophy) has worked as a consultant to several animal protection NGOs and previously as academic associate, working with animal welfare policy and legislation at the Ministry of Justice and later the Ministry of Food, Agriculture and Fisheries. In recent years, she has been collaborating with academic institutions and private sector laboratories to promote the 3Rs and their implementation into research practice. She is particularly interested in behavioural deprivation of animals and has written reports and a peer-reviewed article on the subject.

Helena Röcklinsberg (MTh, PhD in ethics) is an associate professor in ethics and teaches animal ethics at the Swedish University of Agricultural Sciences as well as performing research on ethical issues in relation to animals and food from bachelor to PhD students. Röcklinsberg's research covers ethical issues related to animals used in research and food industry, including fish and insects, as well as those inherent in the veterinary profession. She has been a member of the committee for ethics and education in animal research at the Swedish Board of Agriculture since 2008, followed by the Expert group on animal research under the National committee since its establishment in 2019.

Dorte Bratbo Sørensen (DVM, PhD in ethology and animal welfare) is an associate professor at the Department of Veterinary and Animal Sciences at the University of Copenhagen and researches and teaches animal welfare for laboratory animals on pre- and postgraduate levels. She is the founder of CeLAT, Center of Laboratory Animal Training, a small centre working with improving welfare through handling and husbandry. She has been a member of the Danish Competent Authority for experimental animals since 2009.

Contributors

Cheryl Abbate
Department of Philosophy
University of Nevada
Las Vegas, Nevada, USA

Samuel Camenzind
Department of Philosophy
University of Vienna
Vienna, Austria, APART-GSK Fellow (ÖAW)
and
Messerli Research Institute, University of Veterinary Medicine
University of Medicine
University of Vienna
Vienna, Austria

Josephine Donovan
Department of English University of Maine, Orono, Maine, USA

Mickey Gjerris
Department of Food and Resource Economics
University of Copenhagen
Copenhagen, Denmark

Elisabeth Jonas
Department of Animal Breeding and Genetics, Swedish University of Agricultural Sciences
Uppsala, Sweden
and
Förderverein Bioökonomieforschung
Bonn, Germany

Lisa Kemmerer
Professor Emeritus and Founder of Tapestry, Billings, Montana, USA

Anna Kornum
Independent Researcher
Copenhagen, Denmark

Laura M. Plunkett
Department of Environmental Science, Baylor University
Waco, Texas, USA
and
BioPolicy Solutions LLC
Houston, Texas, USA

Helena Röcklinsberg
Department of Animal Environment and Health, Swedish University of Agricultural Sciences
Uppsala, Sweden

Larisa Rudenko
Program on Emerging Technologies, Massachusetts Institute of Technology
Cambridge, Massachusetts, USA
and
BioPolicy Solutions, LLC
Ventura, California, USA

Dorte Bratbo Sørensen
Department of Veterinary and Animal Sciences
University of Copenhagen
Copenhagen, Denmark

Tatjana Višak
Department of Philosophy
University of Mannheim
Mannheim, Germany

James Yeates
CEO, World Federation for Animals,
Boston, Massachusetts, USA

1

Introduction

MICKEY GJERRIS
University of Copenhagen

ANNA KORNUM
Independent Researcher

HELENA RÖCKLINSBERG
Swedish University of Agricultural Sciences

DORTE BRATBO SØRENSEN
University of Copenhagen

1.1 ETHICAL CONCERNS RELATED TO GM ANIMALS

The use of animals for scientific purposes has always been a sensitive and controversial subject, from the early vivisectionists to the modern-day animal testing in academic and industrial laboratories. Animal-based research presents us with a profound dilemma: on the one hand, it would, to many, be wrong to forego the opportunity to obtain knowledge, which could potentially yield cures or alleviation for serious human and animal diseases and suffering, but on the other hand, the suffering and harming of animals is a source of great ethical and societal concern. From a philosophical point of view, this trade-off, where animals are

DOI: 10.1201/9780429428845-1

sacrificed for 'the greater good', is permissible according to some theories of animal ethics, while unacceptable in others (Röcklinsberg et al., 2017).

On top of the moral concern for the potential pain and suffering of laboratory animals, another ethical dimension has been added since the 1980s: that an increasing number of animals used in laboratories are now genetically modified. The advent of CRISPR and other gene-editing technologies has only added to these issues. In relation to agriculture and genetically modified crops, the tampering with the genetic code has been met with varying degrees of scepticism, especially in Europe (Mielby et al., 2013). This has been due to perceived risks to humans and ecosystems, the socio-economic effects and concerns about the perceived unnaturalness of the technology. The use of biotechnological tools in research animals has mostly attracted attention with regard to the production of humanized, transgenic animals and chimeras. In both cases, the genetic modification of animals may raise questions of our deep-seated notions of 'a natural order' and of human power and identity, possibly connected to the perceived 'unnaturalness' of genetic engineering and a wider unease about science and technology in general (Macnaghten, 2004).

The discussion about the genetic modification of research animals in relation to concepts such as identity, naturalness and integrity as outlined above is one concern. Another group of ethical concerns relate more to risk assessment regarding both the consequences of a modified organism and the organism itself. That is, for one thing, risks of creating potentially harmful organisms such as the unintended activation of dormant retroviruses of other species of animals when, e.g., producing animals suitable for xenotransplantation and for another, the risk of negative welfare implications for the modified animals themselves. These welfare implications can be both intentional and unintentional as the engineering of animals can have both intended and unintended effects at the phenotypical level that are detrimental to the welfare of the animals (Gjerris et al., 2009). Thus, novel technologies might produce unintended effects that cause the animals significant welfare losses. For example, when employing cloning technologies, negative effects such as maternal dystocia, increased maternal morbidity and mortality, hydroallantois, foetal overgrowth, etc. have been described (Vajta & Gjerris, 2006). These examples warrant increased attention when using technologies such as gene editing or genetic modification as animals with novel genotypes might experience pain and suffering that was not intended. The animals may even express altered behaviour patterns, making it difficult to assess, e.g., pain, thus leaving existing welfare protocols insufficient.

The welfare of animals used in science has been a concern for scientists for centuries, and attempts to ameliorate and minimize suffering have led to the formulation of a set of highly influential guiding principles, known as the 3Rs. The 3Rs (Reduction, Replacement and Refinement) developed by Russell and Burch (1959) aim to ensure so-called humane treatment of experimental animals, both for the sake of the animals and to improve experimental measurements. These principles have since 1959 had a growing impact on the principles for use of animals in research globally and been embedded in legislation and official recommendations in, e.g., the EU (Directive 2010/63/EU) and the US

(NRC, 2011). The 3Rs are often depicted as a win-win situation allowing for more humane treatment of animals and better scientific results at the same time. For a full overview of the impact of the 3Rs on legislation and recommendations, see Röcklinsberg et al. (2017, chapter 5).

It is, however, uncertain whether the production of gene-edited or genetically modified animals is even compatible with the traditional guiding principles for humane use of animals in scientific research. Some authors have pointed out that the inherent waste of animals involved in the development of the desired animal models is fundamentally at odds with the principle of Reduction. The 3Rs are based upon the long-term goal that animal testing should eventually be phased out, whereas GM inherently implies breeding of new individuals in the process to achieve, e.g., a certain mouse strain with a specific genotype. Hence, even if refining a method might be the ultimate goal, reduction cannot be strived for during this process, as the modified animals are an integral part of the research (Ferrari, 2006). It has, however, been argued that as long as the individual experiment is deemed relevant and the 3Rs applied to the point where they do not decrease the value of the science, the overall number of animals used is not ethically relevant (Olsson et al., 2012). At the same time though, there has been growing concerns on the number of animals used, which has led to that, e.g., in the EU, all member states are demanded to report the total number of animals used for experimental and other purposes stated in the EU Directive 2010/63/EU on the protection of animals used for scientific purposes.

1.2 A BRIEF HISTORICAL OVERVIEW

Molecular genetics as a scientific discipline emerged in the 1960s, and by the 1970s, functional DNA had been added to the murine genome (Jaenisch & Mintz, 1974). In the 1980s, microinjections of DNA into the murine cells (Gordon & Ruddle, 1981) enabled the targeting of specific genes and the first transgenic mice appeared (Brinster et al., 1981; Costantini & Lacy, 1981). Laboratories at major universities and pharmaceutical and biotech companies began creating their own transgenic and modified strains of mice, and gradually, the technologies were being applied to other animal species, e.g., nematodes (*Caenorhabditis elegans*), fruit flies (*Drosophila melanogaster*), rats, mice, zebrafish, cattle, and pigs. In 1996, Dolly the sheep was the first mammal to be cloned by somatic cell nuclear transfer (Wilmut et al., 1997) and the world began to grasp the seemingly unlimited potential of biotechnology.

Today GM animals are being used in virtually every field of experimental research. Recent technological advances have made the tools (e.g., restriction enzymes and CRISPR/Cas9) for editing genes more precise and cost-effective. A wide selection of modified strains of especially mice and zebrafish are available as objects in research on basic biological studies of gene expression and to model a staggering number of human diseases including, e.g., Huntington's chorea and Alzheimer's. Genetically engineered pigs have been created as large animal models to study and develop therapies for human diseases and as potential transgenic donors for xenotransplantation (Röcklinsberg et al., 2017).

Genetic modification of an animal was early on defined as:

> animals modified either via a technique known as transgenesis (when individual genes from the same or a different species are inserted into another individual) or by the targeting of specific changes in individual genes or chromosomes within a single species – targeted removal of genes (knock-outs) or targeted addition of genes (knock-ins).
>
> *(The Royal Society, 2001:3)*

Another influential definition of modern biotechnology from the World Health Organization (WHO) and the Food and Agricultural Organization of the United Nations (FAO) is:

Modern biotechnology means the application of:

i. In vitro nucleic acid techniques, including recombinant deoxyribonucleic acid (DNA) and direct injection of nucleic acid into cells or organelles; or
ii. fusion of cells beyond the taxonomic family that overcome natural physiological reproductive or recombinant barriers and that are not techniques used in traditional breeding and selection. (WHO & FAO, 2009:2)

It should be noted that as new technologies are developed all the time, such definitions will have to adapt with the technological progress. In Chapter 2, we will provide a more detailed overview of current technologies, and in Chapter 3, we will provide a detailed discussion of current definitions used within regulatory frameworks. Here it suffices to say that genetic modification can involve the deactivation of genes ('knock-out'), the addition of exogenes (either artificial or from another individual) into a living animal, editing the existing genome (e.g., by CRISPR/Cas9) or creating chimeras by mixing the zygotes (thus producing an animal with a genome from two or more organisms). Although cloning is, not strictly speaking, a form of modification, it is still worth including from an ethical point of view as it is subject to the same concerns as modification and can be part of the process for making a desired modified genotype widely available.

1.3 THE RELEVANCE OF THE SUBJECT

Research animal ethics, understood as the task of unfolding the ethical issues related to the use of animals in research, can be seen as a sub-branch of animal ethics that is again a sub-branch of the fields of ethics, a part of philosophical inquiry belonging to the Arts. It can seem like a very narrow subject, but as both the number of animals used and the complexity of the ethical issues show, it is a very relevant field in its own right with connections to many other issues within animal ethics, research ethics and ethics in general.

Not all countries report that they use genetically modified research animals, but the more detailed statistic in the United Kingdom shows that more than 3.06 million scientific procedures have been conducted in 2021 on living animals (Annual Statistics of Scientific Procedures on Living Animals, UK 2021). Among those were 1.33 million animals used in the creation and breeding of genetically altered animals (43% of the total), in the report defined as animals whose genes have mutated or have been mutated (parents of genetically altered animals are not necessarily part of experimental procedures). Among those, most animals involved were mice, followed by zebrafish. However, larger livestock animals such as sheep, cattle or pigs were also used in the procedures, many of which being applied research.

The Directive 2010/63/EU has changed the reporting of animals used for scientific purposes in the European Union, and reports have been published from the end of 2019. The latest report shows that over 10.4 million animals were used for scientific purposes in the European Union and Norway. Around half a million of these animals (508,320) were classified as genetically altered animals to support scientific research, used for the creation of new lines or for the maintenance of existing colonies (Allures Statistical Database).

Research animals, including genetically modified animals, are used in other countries and regions around the globe, but as the above data should be enough to provide an understanding of the relevance of the issues and the problems of finding relevant statistics, we will leave it as this and just end by saying that Taylor and Alvarez (2019) estimate that globally around 192 million animals were used in research in 2015, without specifying how many were genetically modified. An educated guesstimate is that it is a significant amount (Figure 1.1).

Figure 1.1 Three cases of gene-edited animals will be discussed throughout this book, in terms of welfare and from five different ethical positions. Mouse with 'knock-out' of the *Tyr* gene for pigmentation (albinism), zebrafish with induced deletion of the *MMP21* gene (heterotaxy), and rhesus macaque with induced mutation of the *Dystrophin* gene (Duchenne muscular dystrophy). (Illustration Anna Kornum.)

1.4 TIPS FOR READING

There is no general agreement on the technical term to be used when discussing recent technological developments within biotechnology. Genetic manipulation, genetic modification, genetic engineering and gene editing are just some of the terms used. Obviously, there are huge technological differences between different kinds of biotechnology, e.g., transgenesis, cloning and gene editing. Whether these differences are also ethically relevant is a question that can be debated. Such a debate is, for example, seen within the area of plant biotechnology where, whether there is a difference between transgenic and gene-edited plants from a legal and ethical perspective has been the subject of much debate (Röcklinsberg & Gjerris, 2018). In this book, the term "genetic modification" encompasses a range of modern biotechnological tools, including transgenic technologies and gene-editing technologies. The term "genetic engineering" is used in the same fashion, whereas more specific terms such as "gene editing", "cloning" and "knock-out" are used to specify a certain kind of biotechnology. In this book, we prefer to reflect on the reality of the literature and the current societal debates rather than imposing a strict systematic system that will not fit what the reader will encounter elsewhere in the literature. Hence, we have allowed for the co-authors to use the terminology they find appropriate. It should, however, be clear from the context what the terms used cover.

The technological developments and issues described above warrant a continued attention to the ethical, legal and social challenges that the use of animals raises. Some of these will be well-known from the field of animal ethics but take on new shades of grey in the light of the technological possibilities, while others will face us with new choices and value conflicts. This book therefore sets out to explore genetic modification of animals for scientific purposes in the context of the ethical, legal and social issues that the technology raises with a focus on animal welfare and animal ethics. It should be noted that although numerous suggestions on how to distinguish between 'ethics' and 'moral' can be found in the literature, we have chosen to use them interchangeably in this book, whenever not explicitly stated.

To reach this goal, we have worked together with a number of distinguished colleagues to provide information on the state-of-the-art of the technology, public perceptions, legal and regulatory issues as well as a presentation of the most important positions in the ethical landscape of animal research today and how specific cases could be viewed from these positions. Finally, we present a chapter looking into the specific challenges facing approval committees such as Animal Ethics Committees, Institutional Animal Care and Use Committees, etc. and give advice on how to include ethical considerations into the work of evaluating research proposals.

It is not the point of this book to argue a certain viewpoint or ethical position. Our hope is that this book will be a useful tool for students, researchers, administrators, politicians, members of committees that oversee the use of animals in research and the generally interested citizen to gain a better understanding of the ethical issues embedded in the current reality of research animals where novel biotechnological opportunities both accentuates old issues and raises new ones. Our ambition is thus not to tell the readers what the "right" or "correct" ethical opinion is, but rather to help them form a qualified opinion and gain a better understanding of viewpoints that differ from their own.

REFERENCES

Brinster, R. L., Chen, H. Y., Trumbauer, M., Senear, A. W., Warren, R., and R. D. Palmiter. 1981. Somatic expression of herpes thymidine kinase in mice following injection of a fusion gene into eggs. *Cell* 27, 1:223–31.

Costantini, F., and E., Lacy. 1981. Introduction of a rabbit β-globin gene into the mouse germ line. *Nature* 294:92–4.

Ferrari, A. 2006. Genetically modified laboratory animals in the name of the 3Rs? *ALTEX* 23, 4:294–307.

Gjerris, M., Olsson, I. A. S., Lassen, J., and P. Sandøe. 2009. Ethical perspectives on animal biotechnology. In *Handbook of Genetics and Society: Mapping the New Genomic Era. Genetics and Society Book Series*, ed. P. Atkinson, P. Glasner, and M. Lock, 382–98. New York: Routledge.

Gordon, J. W., and F. H. Ruddle. 1981. Integration and stable germ line transmission of genes injected into mouse pronuclei. *Science* 214, 4526:1244–46.

Jaenisch, R., and B. Mintz. 1974. Simian virus 40 DNA sequences in DNA of healthy adult mice derived from preimplantation blastocysts injected with viral DNA. *Proceeding of the National Academy of Sciences of the United States of America* 71, 4:1250–54.

Macnaghten, P. 2004. Animals in their nature: A case study on public attitudes to animals, genetic modification, and 'nature'. *Sociology* 38, 3:33–51.

Mielby, H., Sandøe, P., and J. Lassen. 2013. Multiple aspects of unnaturalness: Are cisgenic crops perceived as being more natural and more acceptable than transgenic crops? *Agriculture and Human Values* 30, 3:471–80.

National Research Council (NRC). 2011. *Guide for the Care and Use of Laboratory Animals: Eighth Edition*. Washington: National Research Council of the National Academies.

Olsson, I. A. S., Franco, N. H., Weary, D. M., and P. Sandøe. 2012. The 3Rs principle – mind the ethical gap! In *ALTEX Proceedings: Proceedings of the 8th World Congress on Alternatives and Animal Use in the Life Sciences, Montreal 2011*, 333–36. Johns Hopkins University Press.

Röcklinsberg, H., and M. Gjerris. 2018. Potato crisps from CRISPR-Cas9 modification - aspects of autonomy and fairness. In *Professionals in Food Chains*, ed. S. Springer, and H. Grimm, 430–35. Wageningen: Wageningen Academic Publishers.

Röcklinsberg, H., Gjerris, M., and I. A. S. Olsson. 2017. *Animal Ethics in Animal Research*. Cambridge: Cambridge University Press.

Russell, W. M. S., and R. L. Burch. 1959. *The Principles of Humane Experimental Technique*. London: Methuen.

Taylor, K., and L. R. Alvarez. 2019. An estimate of the number of animals used for scientific purposes worldwide in 2015. *Alternatives to Laboratory Animals* 47, 5–6:196–213.

The Royal Society. 2001. *The Use of Genetically Modified Animals*. London: The Royal Society.

Vajta, G., and M. Gjerris. 2006. Science and technology of farm animal cloning: State of the art. *Animal Reproduction Science* 92:210–30.

Wilmut, I., Schnieke, A. E., McWhir, J., Kind, A. J., and K. H. Campbell. 1997. Viable offspring derived from fetal and adult mammalian cells. *Nature* 385:810–13.

World Health Organization (WHO) and Food and Agriculture Organization of the United Nations (FAO). 2009. *Codex Alimentarius. Foods Derived from Modern Biotechnology*. Rome: WHO & FAO.

2

The science behind GM research animals

ELISABETH JONAS
Swedish University of Agricultural Sciences
Förderverein Bioökonomieforschung

ANNA KORNUM
Independent Researcher

MICKEY GJERRIS
University of Copenhagen

HELENA RÖCKLINSBERG
Swedish University of Agricultural Sciences

DORTE BRATBO SØRENSEN
University of Copenhagen

DOI: 10.1201/9780429428845-2

2.1 INTRODUCTION

Research animals have contributed to a large part of the scientific knowledge gained up to date. Genetic studies on animals in the laboratory aim to identify and/or further confirm the genetic background of important traits including inherited diseases and to test the effects of defined regions on the genome. Animals can thereby be used as model organisms. Small rodents, especially mice, are often used as models besides the classical model organisms of *Drosophila melanogaster* (Fruit fly) and *Caenorhabditis elegans* (Nematode) or *Danio rerio* (Zebrafish). Typical experiments are knock-outs or knock-ins, which stop or change certain regions on the genome to test the function of genes. There are also experiments to modify defined regions on the genome to test the effect. Also, other animal species are being used as (genetic) models for diseases in humans. Pigs are used as a larger research animal to test their use as organ donors for humans, the so-called xenotransplantation. Many larger animal models exist for diverse human diseases (sheep as model for neuronal ceroid lipofuscinoses (Palmer et al., 2015)), the production of pharmaceutics (anti-thrombin produced in milk from goats (Wang et al., 2014)) or the improvement of agricultural production (AquaAdvantage salmon with faster growth rate (Yaskowiak et al., 2006)).

The aim of many experiments using animals is the identification of consequences of changes in the genome. The genome of individuals contains important information for their development and maintenance and is inherited from parents to offspring. The genome is structured in different chromosomes, the total number of chromosomes often differing between species. Each chromosome contains the DNA (deoxyribonucleic acid), which is closely packed around histone proteins, to fit into the cell nucleus. Information from the DNA is used by mechanisms of the cells to be transcribed into smaller fragments, the RNA and translated into proteins. These processes are happening in individuals continuously to keep the organism functioning. While the processes of translation and transcription are cell specific and thus different information from the DNA are used in different cell types, the general information contained in the DNA is mostly identical across all cells in an organism. Most animals are diploid, which means that they contain two copies of each chromosome and thereby the DNA information. The DNA is built as a double helix and contains four nucleotides: adenine, cytosine, guanine and thymine, which are called bases. Each base is paired with the other, and adenine and thymine as well as cytosine and guanine are connected via sugar molecules (ribose) and phosphate groups.

When talking about the information contained on the DNA, researchers discuss their sequence, which is a long chain of adenine, cytosine, guanine and thymine, or just A, C, G and T. Sections of the base pair chain are further defined as gene. Only a small part of the DNA is actually structured as genes, but it has been suggested that also regions outside gene constructs may play a relevant role for the organism. Nevertheless, most studies, especially those using research animals, focus on actual genes. Genes have defined structures including regulatory elements as the signal for the beginning and end of a gene as well as sections which are used for the translation into proteins (exons) while information on other parts are not translated (introns). Researchers have during the last decades progressed to identify sections on the genome relevant for certain traits and have further identified the function of many genes on the genome. The term "function of a gene" is thereby used for a wide range of definitions. This includes most basically which protein the gene codes for. But it also includes the relevance of the gene or rather the protein it codes for, to the organism. Researchers link genes often to other genes in pathways or cell processes. However, input, for example, results from research, is required to build the pathways, draw the cell processes and finally understand the biological meaning behind a gene.

It has been established that a mutation – which includes a change – in only a single base pair may alter the function of the gene and thus the well-being and appearance of an organism. This change can either be positive, for example, lead to higher resistance to diseases, be neutral such as increased growth, or be negative, for example, lack of production of proteins leading to life-threatening stages. It has been discovered that some base pair changes have such drastic effects that individuals will not survive. Such mutations are called lethal. Research projects aim to identify the genes and thereafter causative mutations leading for example to lethality, to diseases, health problems, production and reproduction differences. Options in livestock breeding are thereafter to select only those individuals with favourable alleles (or base pair combination) for future breeding. However, to be able to select individuals with the most favourable alleles, variation has to be present in the population. This has been done ever since domestication as the selection for certain visible traits or phenotypes has also led to a selection for an underlying genotype. But with the advance of further methods in molecular genetics, it has become possible to identify the underlying base pair combinations and select individuals based on the molecular genetic information.

Genetic modifications can occur in organisms randomly or directed. Random mutations have contributed to the diversity between and within species and are an important part of breeding procedures, for example, in livestock breeding. During the processes of cell growth, the DNA gets replicated many times. There is a likelihood of errors during the replication but especially the meiosis process. However most mutations get corrected instantly, and few remain in the DNA. It is important to take into account that only mutations in the germ cells can be inherited to the next generation. While these errors have contributed to the diversity between and within species, they are random and cannot be predicted. The introduction of mutations by humans has been a research interest for many decades. Artificially introduced mutations have been especially used in plant breeding using mutation breeding, for example, via radiation treatments of plants. As an alternative, directly modifying biotechnologies have been developed since the 1970s. This has firstly been applied in plant breeding to develop plant lines resistant to pathogens and in bacteria for the production of pharmaceuticals such as insulin.

2.2 TECHNOLOGIES

Today two main technologies are being considered by many scientists working with genetic models, the classical genetic modifications as well as genome or gene editing methods. Classical genetic modifications, leading to genetically modified organisms (GMO) are creating transgenes and require the insertion of gene constructs into the DNA. The more recent technologies of genome editing such as CRISPR-Cas9 make use of the cell's self-reparation methods, which allows the change of one or few selected nucleotides at a time. The use of genetic modifications is often a step to gain knowledge on gene functions, and it is therefore an important part of the use of animals in research including research animals. Research animals are used to test different genetic modification, but they are also used to refine technical aspects of the methods for modification. To test gene functions, often smaller model animals such as mice are used, while a number of examples of research livestock exist, which are used to test possible methods for the selection of future breeding animals.

In order to understand the concepts of directed genetic modifications, it is important to understand the technical process and the different methods in the laboratory. One important part of the creation of modifications in the genome is the development of the embryo. The first part is the fertilization of the ovum after which the zygote develops to further 2-cell, 4-cell, 8-cell and 16-cell (morula) stages (Figure 2.1). Thereafter the blastocyst stage has developed, which after hatching from the zona pellucida attaches to the uterine wall during the implementation stage. The genetic transfer is performed during early stages of the embryo development in the lab, outside the female. Unfertilized or fertilized egg cells are flushed or harvested (removed) from a female animal, called donor. Donor animals can be slaughter animals (for example, cows) or living animals. In smaller models such as mice, donor animals are often euthanized for the experiment. Unfertilized egg cells can be fertilized in the laboratory using in vitro fertilization.

The genetic transfer into the developing embryo is usually done in the zygote, or 1-cell stage. This will ensure that the transfer is later present in all cells. The embryo develops further, and the quality of the embryo can be accessed in the laboratory, including the shape and colour. Quality measures will give an indication of the likelihood of the successful development of the embryo. Embryos with a good quality are transferred back into a female individual as early as possible

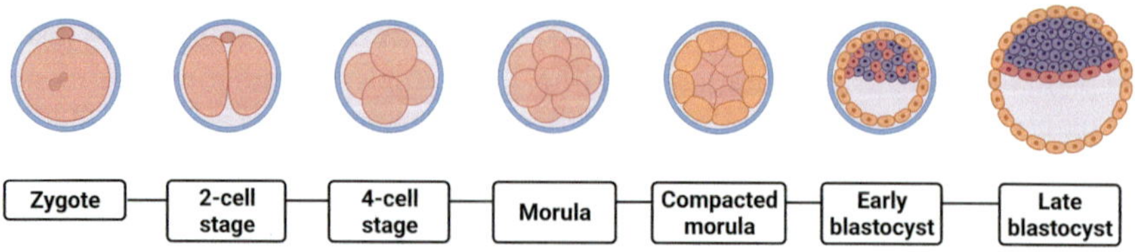

Figure 2.1 Cell stages of development from fertilized egg to blastocyst. (Created with BioRender.com.)

after the genetic transfer but before the implementation stage. It has been shown that human embryos, for example, can develop for up to 2 weeks in a Petri dish, and implement onto the dish (Shahbazi et al., 2016). It is important to note that 14 days was set as a limitation to keep human embryos in a Petri dish by ethicists and scientists already in the 1970s as this is the threshold before the nervous system develops (Ethics Advisory Board DoH, 1979). Embryos in other species develop differently; mice, for example, reach the late blastocyst stage already after 4.5 days compared to 6 days in human (Vázquez-Diez and FitzHarris, 2018). Procedures therefore have to be fast to allow a timely transplantation of modified embryos. Embryo technologies play a critical role for the success of genetic modifications. Research animals are used as donors and recipients of the modified embryos, while being not directly affected by the changes but by the procedures of harvesting and implantation. The other large group of research animals in such research are the offspring developing from the implanted embryos.

Genetic modifications are often separated into the classical methods using a gene construct and genome editing based on the self-repair mechanisms of the DNA in the cell nucleus. Classical genetic modification methods require as a first step to select or make a gene construct. The gene construct contains parts of the DNA or even an entire gene. It can be designed to be inserted at a specific site or randomly into the genome. The second step is to transfer the gene construct into the cells. The last step of the classic genetic modification is the integration of the exogene into the host DNA. The efficiency of this process is very low and off-targets are high. Off-targets are changes to other parts of the DNA strand than the one aimed for. These mutations are therefore nonspecific and unintended, and it is difficult to identify these to ensure that the results are actually achieved by the target mutation.

Genome editing, different from classic genetic modifications, does not require the external gene construct. These technologies make use of enzymes (in the case of genome editing nucleases) allowing site-directed mutagenesis. The techniques of genome editing include oligonucleotide-directed mutagenesis, zinc finger nucleases, transcription activator-like effector nucleases (TALEN), homing endonucleases and clustered regularly interspaced short palindromic repeats (CRISPR) and CRISPR-associated systems (Cas). The last four techniques are also known as site-directed nucleases. These elements can be inserted into the cell, where they introduce a cut. They align to specific sides of the DNA except one or few bases. The self-repair mechanisms of the DNA then attempt to repair the strand, introducing an optional base pair change. As no external DNA is introduced into the cell, only the modifications are introduced by the site-directed nucleases. For oligonucleotide-directed mutagenesis, short DNA fragments with the desired base pair variant are introduced into the cell. This technology is again based on the repair mechanism of the cell, replacing one base in the mismatching base pair.

One important part in both the classic genetic modification and genome editing technologies is the need to insert the gene constructs or the nucleases into the cell. Different methods exist for the transfer into animal cells including pronuclear microinjection, cytoplasmic injection, perivitelline injection using lentiviral vectors, sperm-mediated transfer, electroporation, (somatic) cell-mediated nuclear transfer (SCNT), and transposons or gene targeting using embryonic stem cells (see Figures 2.1–2.4). Success rates and applications of the different technologies are widely species-specific and often low. Pronuclear microinjection is one of the most commonly used methods with success rates between 5% and 30%. It is a mechanic procedure where the zygote or egg cell is held by a pipette while the element is inserted into the pronucleus using a micropipette which penetrates the cell membrane. Also, the injection into the cytoplast is an option for the transfer of nucleases. Nucleases can also be transferred into a cell using electroporation (Figure 2.2).

Alternatively, gene constructs or nucleases can be transferred using sperm (Figure 2.2). Sperm has the ability to bind exogenous DNA for transfer. The external constructs are included in the spermatozoa before fertilization via artificial insemination or in vitro fertilization. One other method is based on the ability of retroviruses to integrate into the genome of a host (Figure 2.2). Lentiviruses, which are part of the family of retroviruses, can be used to transfer the gene construct into the perivitelline space of the cell. This method is usually very efficient with a lower risk of mosaics. Genetic mosaics (having two or more cell lineages with different genotypes) occur after the zygote has developed and they are found to be more likely in in vitro methods. The efficiency of lentiviral vectors is higher compared to microinjection. Retroviral DNA is randomly integrated into the host DNA, a common process happening also naturally for some retroviruses.

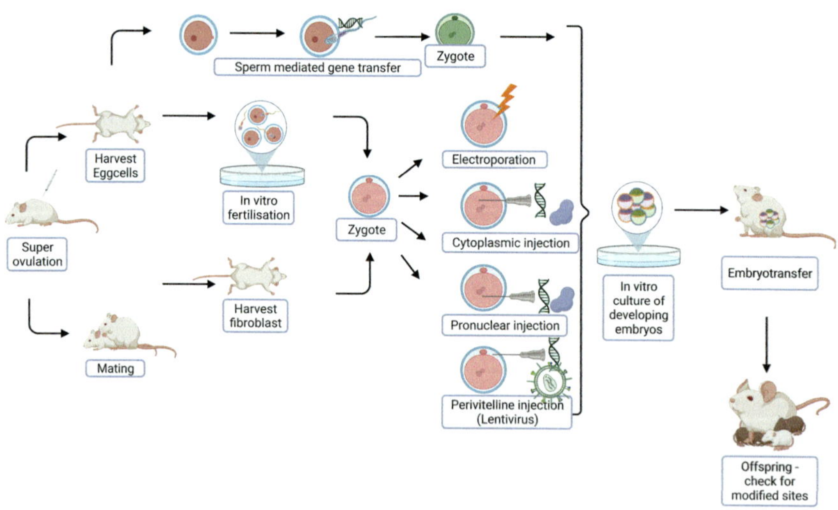

Figure 2.2 Schematic overview of the options for transfer of gene constructs or nucleases in mice. Transfer into oocytes is possible via sperm cells, while transfer into zygotes or in vitro fertilized oocytes can be done using electroporation (of nucleases) during the one-cell stage, cytoplasmic injection (of nucleases), via pronuclear injection (of gene constructs, small RNAs or nucleases) before the merging of the maternal and paternal pronuclei or via perivitelline injection using lentiviruses. (Adapted from Menchaca et al., 2016; created with BioRender.com.)

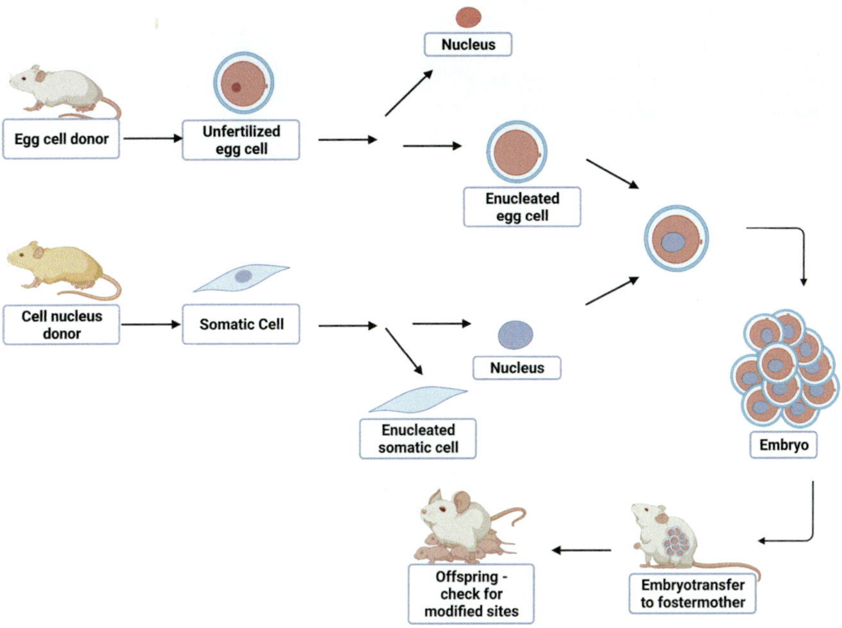

Figure 2.3 Somatic cell nuclear transfer is based on cell cultures which are injected in empty donor cells to be transferred into the cytoplasm via electrofusion. The transfer using embryonic stem cells produced chimeric offspring and is based on cell culture and blastocysts. (Adapted from Menchaca et al., 2016. Created with BioRender.com.)

One other common method is the somatic cell nuclear transfer (Figure 2.3). This method is especially applicable to produce knock-out and knock-in models. It is a common method also applied for cloning. The genome of the oocyte is, in this method, removed from the cell and can be replaced with donor information, for example, genetically engineered information.

And finally, the use of embryonic stem cells has been investigated for gene targeting (Figure 2.4). Embryonic stem cells are pluripotent and have the ability to develop into any tissue, while adult stem cells can only develop into a limited number of cell types. Embryonic stem cells are produced by removing the cells from the blastocyst for culture. They will divide but, unlike in the embryo, not further differentiate into the different cell types. Embryonic stem cells can be further manipulated and applied for the development of individuals. However, this method is currently only used on mice but has not yet been successful in other animal species.

The use of genetically modified research animals aims mostly for the analysis of consequences of small or larger changes in the genome, knock-out, knock-in or silencing of genes. To test the relevance of identified single nucleotide polymorphisms, which are mutations of one or few base pairs or nucleotides in the genome, researchers often use predictive computer models (in silico prediction

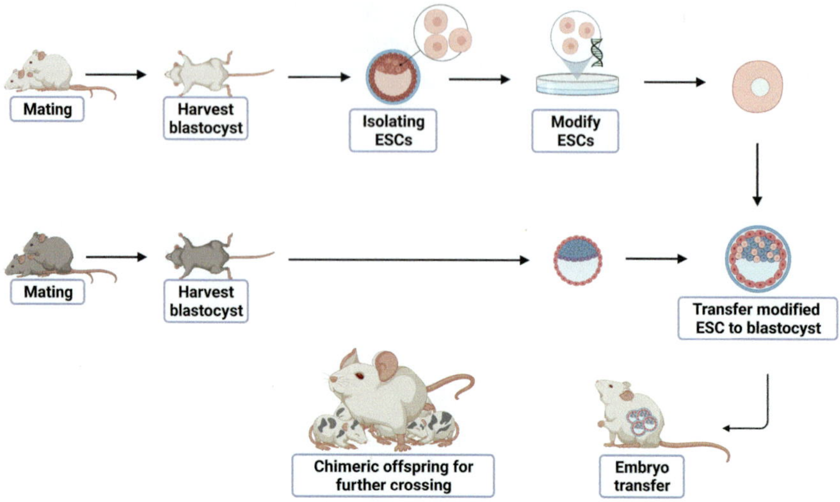

Figure 2.4 The transfer using embryonic stem cells (ESCs) produces chimeric offspring and is based on cell culture and blastocysts. (Created with BioRender.com.)

methods). However, these are not entirely able to predict all underlying consequences of a mutation; models in the laboratory to test the impact of changes in the genome on the organism are therefore necessary for final proof. Genetically modified animals assist to identify consequences of mutations in the population. Targeted integration of a genetic modification is required in knock-out and knock-in models, a determined transgene has to recombine to a specific section of the chromosome. Knock-out of genes has especially been used in mouse models to test the function and dependency of genes.

Another method that, however, allows only temporary silencing or inhibiting of the function of a gene is called RNA interference. Small RNA molecules which are specifically designed to align to defined positions of the genome are transferred into the cell. They attach to the DNA and inhibit the transcription and thereby translation process. However, these small RNA molecules are temporary, and the genes are silenced only for a short time.

2.2.1 Success rates

Success rates for classic modifications are relatively high in mice with 15%–20% of the transferred mice surviving until birth, of which 4%–5% are transgenic. These numbers are lower in pigs (0.2%), sheep, goats and cattle (0.7%) (see Ferrari, 2006). Genome editing methods are leading to higher success rates of the genetic change and have to some degree replaced the more classical methods. It is further relevant for both the classic as well as genome editing methods that not all transgenic individuals will be able to inherit the change to the next generation and a larger proportion are heterozygote, which means that they are only carrying one of the copies of the transgene. Looking into the next generation of animals, a carrier with non-carrier mating will only lead to 25% offspring carrying one copy of the transgene (carrier). If two carrier are mated, this will result in 50% carriers but only 25% homozygote transgenic individuals. Mating of a non-carrier with a homozygote transgenic individual will result in all offspring carriers, while only mating of two homozygote transgenic individuals will ensure that all offspring are homozygote for the transgene. It has been stated that homozygote animals are required to establish a transgenic line of research animals. The number of individuals required for an experiment in a PhD project using classical modifications was thereby predicted between 300 and 4,000 animals (de Cock Buning, 2004). Recent statistics confirm this tendency, for instance.

An experimental design using research mice for transgenic experiments requires a large number of mice. With 15–20 mice, 200 zygotes can be collected after hormonal superovulation. If 95% of these succeed the microinjection, 190 zygotes can be implanted. If 30–40 zygotes are implanted in each mouse, around five female recipient mice are required. If 25% of those survive, it will lead to around 40–50 surviving mice (around ten per mouse), of which 10%–20% can be used as founder for the transgenic line. It is important that in the described experimental setting, all female donor mice are usually euthanized in order to collect the zygotes and male mice must be replaced regularly. Regarding the length of the procedures, microinjection takes around 2 hours for 200 zygotes and the surgery to implant the zygotes takes around 3 hours for all mice (Cho et al., 2009). These procedures differ in larger animal models, for which less invasive methods are used for both harvesting of zygotes and implantation of injected zygotes.

2.3 RELEVANCE OF RESEARCH ANIMALS

Genetically modified research animals are considered an important part of research. They can be used to study the effects of genetic modifications in more detail, to understand gene functions and interactions, to develop diagnostic methods, to test the outcome of possible treatments or even study options for xenotransplantation. Natural occurring mutations offer one possibility and a number of examples of research animals based on such mutations exist (Andersson, 2016; Pinnapureddy et al., 2015). There are many natural mutations, which have been or have the potential to be the basis for research lines of animals in the future. These include naturally occurring mutations in animals (https://www.omia.org/home/) or mice (http://www.informatics.jax.org/) (Baldarelli et al., 2021). Examples include various phenotypes of disease models, developmental deficits with more or less consequences for the individual. However, the number of animals with natural mutations is limited as most individuals will be selected early on or die. Technologies offering the introduction of mutations have offered new opportunities to use animals as models. Genetic modification in mice (Brinster et al., 1981; Gordon et al., 1980) initiated the beginning of the application of this technology in animals. They have been used as transgenic models for the study of diseases in humans. Mouse models have also been successfully used to study the functions of genes. The first transgenic mice were produced using directed mutations. Genetically modified mice are used as models to study genetic disorders, multifactorial and polygenic (complex) disorders, inflammatory diseases, neurodegenerative diseases and cancer in humans.

In 1985 the first successful modifications of rabbit, sheep and pigs were performed via microinjection. A number of examples of genetically modified livestock in research exist. Livestock models have been suggested for organ donation, disease models or bioreactors. Pigs in the laboratory are used for the study of xenotransplantation and also as models for different diseases in humans including Alzheimer's disease, Huntington's disease, spinal muscular atrophy, retinitis pigmentosa, macular dystrophy, cardiovascular diseases, cystic fibrosis, diabetes mellitus, liver disease, osteoporosis, skin disease, cancer, infectious diseases or immunodeficiency. Most of these models are designed using nuclear transfer and gene targeting or random transgene placement. There had been approaches of chimeric pigs (chimeric is defined as an individual carrying cells with different DNA information, for example, transplants) as models for organ transplantation as xenotransplantation is still an important biomedical topic. Many research pigs as models for traits related to xenotransplantation exist (Cooper et al., 2016).

Genetically modified fish exist since the first trials in 1984 in rainbow trout, and most fish were produced using microinjection. Some fish species are especially interesting models for further research of cell mechanisms including development and thus improvements of techniques as embryos are nearly transparent. Fish are also used as models for environmental experiments as variation regarding cold tolerance or different measures of water surroundings exist. Genetic modifications in avian species are more challenging due to oviparity and the properties of the ovum. Genetically modified chicken can be used as avian disease resistance models, bioreactor models and human disease models. The zebra finch has been suggested as an interesting and novel model for different human diseases, for which no other good animal models exist. It has been suggested that the further development and application of gene editing methods in avian will lead to more gene-specific avian knock-out models, allergen-free poultry, human disease models, egg-based bioreactors or an avian disease resistance model.

One other application of genetically modified research animals, especially livestock, is application for altered production and disease resistance or tolerance in farm animals. The aim of the first genetically modified farm animal which has been approved for human consumption in some countries has an altered growth. The AquAdvantage salmon shows faster growth and was developed by classic genetic modification. Another early application of classic genetic modification is the Enviropig. This pig had been modified to be able to break down phosphorus using the enzyme phytase. This resulted in a significant lower phosphor content of the faeces, which was said to have a large positive impact on the environment. However, the population was research only and never used for applied breeding programmes. Other examples of genetically modified livestock exist, but all except the salmon remain currently solely research projects.

2.4 POSSIBILITIES IN THE FUTURE

The opinion of parts of the society towards the use of animals in research has changed; today the use of animals for the research of cosmetics is already forbidden in many countries, as has the use of primates in experiments decreased over the years also as a consequence of public pressure, and more alternatives to the use of animals are therefore needed. Some alternatives to the use of genetically modified research animals have been suggested. These include in vivo, in vitro or in silico methods. In vivo methods would perform experiments on lower order animals. Fruit flies could be an alternative to other animals or vertebrates such as zebrafish in early developmental stages. Such models might be an option to test the functions of some genes; however, they would still require experiments using animals. In vitro methods on the other hand require only few living animals. They offer also the option of using slaughter animals. Testing of the development of early-stage embryos is already today performed using oocytes from slaughter animals. But the question of how early these should be considered as animals will be ongoing. Future research findings will contribute to the discussion among and between research disciplines on the time point an individual should be considered as alive. Such discussions will affect the way animal embryos can be used for research purposes including methods of genetic modification. For other experiments, cell lines are already today commonly used during the first steps of experiments for genetic modification. However, results from such experiments cannot show the entire picture and thus the final consequences of modifications. They further require some tissue from animals.

A third alternative are in silico methods. Computer simulations and machine learning methods are being greatly advanced and already today allow to artificially predict biologic processes and their outcome. However, in silico methods do require reliable input, which will often require input from animal experiments. The mechanisms in the genome are complex and even though a lot of information has been collected, many unknown relationships and dependencies exist, which cannot be predicted in silico or proven in vitro. The use of research animals is therefore often required as final proof of the function of certain modifications in the genome. However, the advancement of personalized medicine, including collection of huge amounts of data using information from genome, transcriptome, microbiome, metabolome and options offered by the 'omics' approaches, also leads to immense datasets which may feed into such in silico methods. If computational options exist to use algorithm, such datasets may allow fairly valuable predictions which can replace a large number of animal models (suggested for cosmetic testing in Fentem et al., 2005).

While the 3R principle plays an important role (see also Chapter 3), the relative simplicity of using the new genome editing technologies allows a wider range of research groups to apply this method and potentially the use of research animals. The currently stated simplicity as well as greater success rates of the genome editing methods might open the use of such methodologies in more regions of the world. This might have an impact on the total number of genetically modified research animals. These methodologies might invite more researchers to test their findings, for example, identify genetic loci associated with certain traits of interest. The more recently developed methods of genome editing are easier and more successful regarding the development of genetically modified individuals. This offers new opportunities for many research labs to put their results on a new level of evidence and application. Previous technical constraints of testing genetic modifications are partly diminished and more researchers aim to use research animals for final proof of their findings. Looking at the amount of genome-wide association studies published, all of which aim to identify the genetic background of inherited traits, it is obvious that a large number of potential loci for testing of genetic consequences are offered. This can also be observed in livestock. A database of quantitative trait loci, regions on the genome showing some statistic relation to traits of interest, presents the large amount of data and identified regions for the main traits. If methods are easily accessible and relatively inexpensive, it might be possible that more results from such association studies will be translated into animal models to test the consequences of identified genomic regions.

One important point in the future will be the clear definition of the technologies of genetic modification especially for the legislation. The definition of genetic modification is being widely discussed. The Cambridge Dictionary defines it as "the process of changing the structure of the genes of a living thing in order to make it healthier, stronger, or more useful to humans" or "the process of scientifically changing the genes of plants and animals in order to create forms of them that are less likely to get diseases, will grow faster, etc." Definitions of genetic modifications are often closely linked to the definition of genetically modified organisms (GMO). While many researchers strictly separate the classic modifications and genome editing, a distinct treatment of both technologies is not agreed by all. While the legislation for research animals has more options compared to that of animals for food production, a strict policy for the production of genetically modified animals for food production will likely also affect the further use of novel techniques such as genome editing in research animals. If new technologies such as genome editing are considered as GMO, it will likely diminish some of the research using genetically modified animals. Such legislations may also likely restrict the dissimilation of genome-edited livestock for food production into the food chain. This will lower the interest to test changes relevant for livestock populations using genetically modified animal models. There might be fewer approvals of experiments if there is no future applicability. While livestock species might still be used as models in the laboratory, experiments using farm animals for future breeding based on genome editing might be less likely. While this will include a number of experiments using livestock, such models display only a smaller proportion of the currently used genetically modified research animals.

Another issue in the future will be to identify animals modified using genome editing as these technologies, unlike classical methods, do not leave a trace in the genome and are not to differentiate by spontaneous mutations. However, it might be important, especially for research animals, to clearly trace the changes made and consequences occurring spontaneously in the genome.

There are further points which might change the technical terms of the use of research animals in research on genetic modifications. In human embryo research, the 14-day rule is currently under discussion (Chan, 2018). This might have an affect also on animal research. Also, human-animal chimers (Yamaguchi et al., 2017) are being widely discussed. This might have an impact on both research on human and also animal embryos. The field of genetic modifications is growing, and scientists seek new options to research the genetic background of development and functioning of organisms. Methodologies for modifications will likely develop further, allowing even more precise changes and especially the reduction of the total number of animals or early developmental stage embryos required in an experiment. Injections with little or no side effects might be the future tool for genetic modifications, allowing an even easier application of such methods for testing the consequences of genetic modifications. Especially the need for complex biotechnological methods during the process of transfer will be possibly dismissed in the future and alternatives will be developed which do not require steps such as cloning and thereby reduce the total number of animals required for a study.

Furthermore, it is likely that more and more alternative methods need to be applied, as the use of animals is already today questioned by a growing proportion of the society. New opportunities might be given by artificial intelligence and machine learning. This might allow a significant reduction of animals used in experiments as such methods will allow better predictions of even biological complex mechanisms. Such methods might be shown in the future to be able to mirror very complex biological mechanisms and as such allow better predictions than those from any model organism.

REFERENCES

Andersson, L. 2016. Domestic animals as models for biomedical research. *Upsala Journal of Medical Science* 121, 1:1–11.

Baldarelli, R. M., Smith, C. M., Finger, J. H., et al. 2021. The mouse Gene Expression Database (GXD): 2021 update. *Nucleic Acid Research* 49, D1:D924–D931.

Brinster, R.L., Chen, H.Y., Trumbauer, M., Senear, A.W., Warren, R., and R.D. Palmiter. 1981. Somatic expression of herpes thymidine kinase in mice following injection of a fusion gene into eggs. *Cell* 27, 1:223–231.

Chan, S. 2018. How and why to replace the 14-day rule. *Current Stem Cell Reports* 4, 3:228–234.

Cho, A., Haruyama, N., and A. B. Kulkarni. 2009. Generation of transgenic mice. *Current Protocols in Cell Biology*. Chapter 19: Unit-19.11.

Cooper, D. K. C., Ekser, B., Ramsoondar, J., Phelps, C., and D. Ayares. 2016. The role of genetically engineered pigs in xenotransplantation research. *The Journal of Pathology* 238, 2:288–299.

de Cock Buning, T. 2004. Practical difficulties in balancing harms and benefits in the modern use of laboratory animals: Biotechnology. *Alternatives to Laboratory Animals: ATLA* 32, 1:459–463.

Ethics Advisory Board DoH. 1979. Education and welfare. Report and conclusions: HEW support of research involving human in vitro fertilization and embryo transfer. Washington, DC: Department of Health, Education and Welfare.

Fentem, J., Chamberlain, M., and B. Sangster. 2005. The feasibility of replacing animal testing for assessing consumer safety: A suggested future direction. *Alternatives to Laboratory Animals: ATLA* 32, 6:617–623.

Ferrari, A. 2006. Genetically modified laboratory animals in the name of the 3Rs? *ALTEX* 23, 4:294–307.

Gordon, J. W., Scangos, G. A., Plotkin, D. J., Barbosa, J. A., and F. H. Ruddle. 1980. Genetic transformation of mouse embryos by microinjection of purified DNA. *Proceedings of the National Academy of Sciences of the United States of America* 77, 12:7380–7384.

Menchaca, A., Anegon, I., Whitelaw, C. B. A., Baldassarre, H., and M. Crispo. 2016. New insights and current tools for genetically engineered (GE) sheep and goats. *Theriogenology* 86, 1:160–169. doi:10.1016/j.theriogenology.2016.04.028.

Palmer, D.N., Neverman, N. J., Chen, J.Z., et al. 2015. Recent studies of ovine neuronal ceroid lipofuscinoses from BARN, the Batten Animal research Network. *Biochimica et Biophysica Acta (BBA) – Molecular Basis of Disease* 1852, 10:2279–2286.

Pinnapureddy, A. R., Stayner, C., McEwan, J., Baddeley, O., Forman, J., and M. R. Eccles. 2015. Large animal models of rare genetic disorders: sheep as phenotypically relevant models of human genetic disease. *Orphanet Journal of Rare Diseases* 10:107.

Shahbazi, M. N., Jedrusik, A., Vuoristo, S., et al. 2016. Self-organization of the human embryo in the absence of maternal tissues. *Nature Cell Biology* 18:700–708.

Vázquez-Diez, C., and G. FitzHarris. 2018. Causes and consequences of chromosome segregation error in preimplantation embryos. *Reproduction* 155, 1:R63–R76.

Wang, C., Huang, Y., Kong, Y., Luo, J., Zhang, G., Zhao, D., Su, Z., and G. Ma. 2014. [Purification of recombinant human antithrombin III expressed in a goat mammary bioreactor] *Chinese Journal of Biotechnology*, 30, 2: 1634–1638.

Yamaguchi, T., Sato, H., Kato-Itoh, M., et al. 2017. Interspecies organogenesis generates autologous functional islets) Interspecies organogenesis generates autologous functional islets. *Nature* 542:191–196.

Yaskowiak, E.S., Shears, M.A., Agarwal-Mawal, A. and G.L. Fletcher. 2006. Characterization and multi-generational stability of the growth hormone transgene (EO-1α) responsible for enhanced growth rates in Atlantic Salmon. *Transgenic Research* 15: 465–480.

3

An overview of the regulation of genetically altered animals in research

LARISA RUDENKO
Massachusetts Institute of Technology
BioPolicy Solutions, LLC

LAURA M. PLUNKETT
Baylor University
BioPolicy Solutions LLC

ANNA KORNUM
Independent Researcher

HELENA RÖCKLINSBERG
Swedish University of Agricultural Sciences

DORTE BRATBO SØRENSEN
University of Copenhagen

MICKEY GJERRIS
University of Copenhagen

DOI: 10.1201/9780429428845-3

3.1 INTRODUCTION

The lexicon used to refer to the regulation, governance, and products of "modern biotechnology" permeates the scientific, legal, and political discourse that applies to what have often been referred to as "genetically modified organisms" (GMOs). The aim of this chapter is to present existing legal and regulatory structures and some of the overarching governance practices that apply to "genetically modified" animals in what are considered "research" settings in selected countries or geopolitical units around the world. As the issue of ethics in animal research has been extensively covered by Röcklingsberg, Gjerris, and Olsson (2017), we incorporate their findings by reference and concentrate on the statutory regulation and softer governance systems in situations in which animals that are clones (or their offspring) or that have intentional genomic alterations that fit within the rubric of "techniques of modern biotechnology" as defined by the WHO and FAO (2009) are used for research purposes (see chapter 1.2).

We expressly expand from the kinds of animals traditionally considered as "laboratory animals" such as mice, rats, rabbits, and zebrafish, to large animals traditionally considered as livestock that have been adopted as large animal models of disease, and to a lesser degree, non-human primates (NHP). In general, regulatory and governance mechanisms have focused largely on risks to the target animals (*i.e.*, the animals that are part of research protocols, including *verum* and untreated controls). Recently, there has been an increasing emphasis on the impacts that either contained or unfettered release of such animals might have on their surrounding environments, including non-target animals, food webs, and geophysical conditions. In general, these topics are beyond the scope of this chapter.

Although the issue of what constitutes an "animal" appears to be straightforward colloquially it is influenced in profound ways by nuance. As a result, before describing the regulatory frameworks that apply to "genetically altered animals" worldwide, a brief discussion of some of these concepts can help to frame the presentation of applicable regulatory frameworks and the broader concept of governance. These concepts include definition of the term "animal", the meaning of the term "modified", the bounds of "research", the distinction between governance and regulation, and the distinction between process-based and product-based regulatory systems.

3.2 WHAT IS AN ANIMAL?

Although the focus of this chapter is on animals used for research purposes, the historical context around the definition of the term "animal" is important for understanding the development of regulatory approaches developed around the world for the use of animals in research.

Early philosophers attempted to define the term "animal", if by no other means than exclusion from the concept of "human". Plato has been cited as saying that "man...is a tame or civilized animal who was also bipedal and featherless" (Plato, 2008). Perhaps apocryphally, he was mocked by Diogenes, who plucked a chicken and thrust it at Plato in his Academy saying "Behold! Here is Plato's man" (Diogenes, 1959). The Academy is reported to have responded by amending Plato's definition by adding "having broad nails". Recounting a myth, Plato further suggested that in a previous golden age, humans and animals coexisted peacefully, with neither killing the other, and that they may even have conversed (Bell and Nass, 2015). This distinction was not universally held, even contemporaneously, as Aristotle defined "human" as "the rational animal", thus rejecting the possibility that any other species is rational (Mesaros, 2014).

Perhaps from this beginning, the concept of humans as "sentient beings" began to permeate the collective consciousness of the Western world. There is some consensus regarding the distinction between humans and non-human animals, as seen in the regulatory definitions in Table 3.1, except with respect to whether a degree of sentience should be expressly considered in the types of oversight various regulatory and governance bodies exercise over certain types of non-human animals.

There are varying degrees to which sentience, consciousness, and rationality enter into categorizations of animals that the cultural norms of modern humans have adopted and refined over the years. These categorizations often have driven the adoption of certain rules and regulations for use of animals in research. Among these are that animals are (1) entities used for food, fibre, transportation, mechanical labour, or protection, either by capture or breeding; (2) companions akin to family members; and (3) surrogates for humans when exposures or procedures are deemed "too immoral or unethical" for humans, *e.g.*, laboratory or research animals. These categories may help direct the evaluation of regulation, governance, and the consequent concepts of ethical standards of care between and among different countries or geopolitical units.

Table 3.1 As can be readily seen, there is no common international agreement between different national legislative definitions.

Country/ geopolitical unit	Definition	Exclusions	Is "sentience" included?
European Union	Defined in terms of category of use (laboratory, farm, companion, *etc.*)	Humans	Yes (since 2009)
Middle East			
Israel	Defined in terms of category of use (research, farm, companion, *etc.*)	Humans	No
North America			
Canada	Defined in terms of category of use (laboratory, farm, companion, *etc.*)	Humans and for GE animals and fish	No
United States	FDA does not define animal, except "man or other animals" in law USDA and NIH include only certain species in their oversight of animals	Humans Certain species are excluded depending on the USDA or NIH regulation/ law	No
Mexico	Defined in terms of category of use (research, farm, companion, *etc.*)	Humans	No[a]

(*Continued*)

Table 3.1 (*Continued*) As can be readily seen, there is no common international agreement between different national legislative definitions.

Country/ geopolitical unit	Definition	Exclusions	Is "sentience" included?
South America			
Argentina	Not defined specifically	Humans	No
Brazil	Live, non-human vertebrates	Humans, invertebrates, and any other species outside of vertebrates (molluscs, arthropods, *etc.*)	No
Chile	Defined in terms of category of use (research, farm, companion, *etc.*)	Humans	Yes (since 2009)
Far East			
Japan	Welfare laws list specific species under the definition of animal (cattle, horses, pigs, cats, dogs, etc.) GM animal law uses the term "organism", not animal.	Humans	No
China	Defined in terms of category of use (research, farm, companion, *etc.*)	Humans	No
South Korea	Welfare laws define animal as vertebrates	Humans	No
India	GM and welfare laws define animal as a living creature	Humans and microorganisms	No
Oceania			
Australia	Federal GM law defines animal as every kind of organism in the animal kingdom including non-vertebrates Welfare laws can differ in the definition by state or territory although most define animal as a live vertebrate species	Humans	No

(*Continued*)

Table 3.1 (*Continued*) As can be readily seen, there is no common international agreement between different national legislative definitions.

Country/ geopolitical unit	Definition	Exclusions	Is "sentience" included?
New Zealand	Welfare laws define animal as any live member of the animal kingdom Law defines a GMO as any plant, animal, insect, or microorganism	Humans	Yes (since 2015)
Singapore	Welfare law defines animal as any mammal or fish other than a human being	Humans	No
Malaysia	Welfare law defines animal as any living creature other than a human being GM law defines animal as a living modified organism	Humans	No

[a] One Mexican state, Michoacán de Ocampo, considers any animal to be "sentient"; no such consideration in Mexican federal law.

As Table 3.1 indicates, most geopolitical units make the first distinction as to what is an animal by separating humans from all other members of the Kingdom Animalia, generally for regulatory but not always philosophical purposes. The definitions in Table 3.1 were developed to fit within legal/regulatory frameworks and therefore can be quite different due to the context of the use of the term "animal". In fact, different laws or regulations may apply to wider or narrower groupings of "animals", even within the same country.

The United States (US) laws and regulations illustrate the definitional differences that exist within one federal government. For example, the US Food and Drug Administration (FDA) does not define "animal" *per se* but does make the distinction between humans and "other animals" (21 U.S.C. 321(v)). In contrast, the US Department of Agriculture (USDA) uses different criteria depending on which enabling statute (law) it employs to exercise oversight. Under the terms of the 1996 Animal Welfare Act (AWA), which is administered by the Animal and Plant Health Inspection Service (APHIS), "animal" is defined as "*any live or dead dog, cat, nonhuman primate, guinea pig, hamster, rabbit, or any other warm blooded animal, which is being used, or is intended for use for research, teaching, testing, experimentation, or exhibition purposes, or as a pet*" (Animal Welfare Act §2132(g)). For the purposes of regulation, the AWA specifically excludes farm animals used for food or fibre; cold-blooded species (amphibians and reptiles); horses not used for research purposes; fish; invertebrates such as crustaceans, insects, etc.; birds; rats of the genus *Rattus*; and mice of the genus *Mus* that are bred for use in research. Birds other than those bred for research are covered under the AWA, and a proposed rule was published in the Federal Register on February 22, 2022 (APHIS, 2022).[1]

APHIS also confines the term "livestock" § 780.120 "*to the ordinary use of the word and includes only domestic animals ordinarily raised or used on farms. It does not include such animals as albino and other rats, mice, guinea pigs, and hamsters, which are ordinarily used by laboratories for research purposes...fish are not 'livestock'....This term excludes: birds, rats of the genus Rattus and mice of the genus Mus bred for use in research...*" (Title 9, volume 1 section 11). As previously mentioned, this law is implemented by USDA and the *Public Health Service (PHS) Policy on Humane Care and Use of Laboratory Animals* (Office of Laboratory Animal Welfare, 2015). Thus, even the first-order definition of "animal" differs not only from country to country but even among different regulatory authorities within the same country, usually in response to the laws.

3.3 WHAT DOES "MODIFIED/ALTERED" MEAN?

Different geopolitical units may use different terms to refer to an animal whose genome has been intentionally altered with the tools of modern biotechnology. In this chapter, we use the acronym set forth by the enabling statutes or other regulatory decrees by the geopolitical unit being discussed, with the understanding that what could be categorized as "science-based" nomenclature can lead to problematic socio-dynamic issues such as regulation and barriers to trade. Genome/base editing is the most recent development in modern biotechnology and has been treated as "GMO" in some regulatory policies (Friedrichs et al., 2019). When not specifically addressing the nomenclature linked to a geopolitical unit, we have used the broader term of "genetically altered".

Perhaps the most commonly used term with respect to legal/regulatory and colloquial usage is "genetically modified animal" (GM animal) or "genetically modified organism" (GMO). Many of its antecedents reside in the *Convention of Biological Diversity* (CBD) which called for a protocol on the safe transfer, handling, and use of Living Modified Organisms (LMO) (the Cartagena Protocol; Secretariat of the Convention on Biological Diversity, 2000). A LMO is defined as "*any living organism that possesses a novel combination of genetic material obtained through the use of modern biotechnology*"[2]; the organism component of that acronym refers to "*any biological entity capable of transferring or replicating genetic material, including sterile organisms, viruses, and viroids*". Thus, under the terms of the Cartagena Protocol, genetically altered animals are referred to as "Living Modified Organisms" or LMOs (Secretariat of the Convention on Biological Diversity, 2000), although the term GM animal or GMO is sometimes substituted.

The European Union (EU) has a prescriptive definition of what constitutes a genetically altered animal, which they refer to as a GM animal:

> The phrase "genetically modified" or GM animal is used when genetic material has been altered by adding, changing or removing certain DNA sequences in a way that does not occur naturally ... DNA is the genetic material of an organism and carries the instructions for all the characteristics that an organism inherits. Changes introduced in an animal's genetic make-up can therefore be transmitted to the next generation.[3]

Other countries, such as the US and Canada, use the phrase "genetically engineered" (GE) animal or "animals with intentionally altered DNA".[4] Regulators in the US have taken the position that genetic modification by humans has been occurring since the time of selective breeding of plants or animals, and that there are very few organisms that have not been "modified" with the exception of wild game, certain kelp and seaweed,[5] *etc.* In the US, genetically altered organisms that are referred to as "GE" are thus considered to be a sub-set of genetically modified organisms. Canada uses both "genetically engineered animal" and "genetically modified animal".

Further, regulatory systems in different countries may differ in what serve as "regulatory triggers", *i.e.*, the characteristic that determines whether, or how, something is regulated. There tend to be two basic approaches: *process*, in which the technology is the trigger for regulation (*e.g.*, cloning or genetic modification); and *product*, where the regulatory trigger is often the statutory definition of the regulated article (*e.g.*, in the US, "*an article intended to affect the structure or function of the body of man or other animals*" regardless of how it is produced). This difference in approach is not insignificant, particularly when new techniques in biotechnology emerge and existing regulatory approaches may need to be amended for process-based regulation. The following discussion on how to regulate genome editing is an example of process versus product regulation.

From a strictly scientific perspective, genome editing can be thought of as a more precise form of modern biotechnology or genetic alteration. By using highly precise nucleoprotein complexes, often referred to as "site-directed nucleases" such as transcription activator-like effector nucleases (TALENs), zinc finger nucleases, and Clustered Regularly Interspaced Short Palindromic Repeats (CRISPR), and relying on a cell's DNA repair capabilities, it is possible to overcome the random genomic integration that is characteristic of more traditional genetic alteration methods. That random integration into the genome gave rise to concerns referred to as "pleiotropic" effects that formed a large part of the assessments of the health of animals (or plants) and the potential risks associated with food from those organisms.[6] Much of the traditional genetic alterations associated with animals have relied on the insertion of genetic sequences from the same or different species in order to obtain a desired phenotype (Garas et al., 2014). Less frequently, molecular biologists were able to effect deletions by relying on recombinational events (Klymiuk et al., 2010).

With the adoption of genome editing, three molecular changes could result in the genome. The first of these alterations is very small deletions or substitutions in the primary sequence of the DNA comprising a genome. The second is the ability to delete large stretches of DNA very precisely; and the third is the ability to insert DNA sequences at very specific sites. The ability to insert DNA sequences is independent of the source of the DNA; it could come from a closely related animal, it could come from an organism that is not closely related to the target animal, or it could be a completely synthetic sequence. Considered from that perspective, all forms of genome editing meet many of the definitions that serve as triggers for the different regulatory regimes. Another categorization of the types of genome editing has been summarized in the OECD report (Friedrichs et al., 2019). These are referred to as site-directed nuclease SDN1, in which there is a site-directed mutagenesis event that is not template driven; as SDN2, in which nucleic acid templates guide the editing; and as SDN3, in which stretches of DNA are inserted in a site-specific manner. The OECD considers the latter two to be "LMOs", using the language of the Cartagena Protocol on Biosafety of the Convention on Biological Diversity.

The EU, which employs a precautionary approach to any type of genetic alteration, proposed to the European Court of Justice that genome editing could be considered as a form of mutagenesis that is not regulated under the GMO Directives (18/20001/EC). The Court ruled that genome editing constitutes "genetic modification" and, at least in the EU, should be regulated under the GMO Directive.[7] In the US, where the regulatory framework is product-based, genetic engineering and genome editing are not considered to be regulatory triggers; the FDA has issued a draft updated Guidance for Industry (January 2017)[8] on animals with intentionally altered DNA clarifying that position. Other geopolitical bodies are struggling with the extent to which any type of genome editing, apart from transgenesis, should be regulated as "GMO". As pointed out at an OECD conference on this issue, "*each one of these types [of genome editing] poses specific challenges to the regulatory considerations pertaining to it and could thus induce technique-specific discrepancies in the relevant governance approaches*" (Friedrichs et al., 2019).

In some regulatory arenas, however, there appears to be a desire among technology developers and animal breeders to avoid what is perceived to be an overly onerous regulatory system. This has resulted in reframing the molecular mechanisms into categories that could lead to different types of regulation, especially for agricultural purposes. Several groups have argued that all genome editing, including SDN3 insertions, that does not involve the transfer of genes from non-closely related sources (transgenesis) should be considered as equivalent to "what happens in nature". They also suggest allele transfer that could be possible as the result of conventional breeding and selection should not be subject to regulation under various genetic engineering/GMO regimes, with either no or much less stringent regulatory oversight (Murray and Maga, 2016; van Eenennaam, 2018; van Eenennaam et al., 2019). Interestingly, for purposes of regulation, the US FDA recently opined that some intentional alterations are virtually identical to those found in nature and has classified them as "low risk".[9]

3.4 WHAT IS "RESEARCH"?

Three broad areas of animal "research" have been recognized: (1) research intended to benefit humans (*e.g.*, toxicology and pharmacology studies for purposes of research and development with the intent of regulatory approval of the human-intended product (*e.g.*, drug, biologic, food)); (2) research intended to benefit humans and that indirectly may benefit animals (*e.g.*, studies on zoonotic diseases); and (3) research intended to benefit animals primarily (*e.g.*, toxicology and pharmacology for animal drugs, conservation studies).

In the area of research that benefits humans, either directly or indirectly, studies have focused on understanding life processes whose primary beneficiaries may be humans, and typically have involved use of rodents and other animals traditionally considered "laboratory" animals (*e.g.*, zebrafish). Other animals (*e.g.*, dogs) have been employed as surrogates for humans in studies of anatomy and physiology. Pigs and NHP are used as sources of cells, tissues, or organs for xenotransplantation, and as test subjects in toxicology studies to assess safety or efficacy of products primarily intended for human use. Such studies are often parts of packages of premarket or preclinical studies for development of drugs, cosmetics, biologics, food additives, or medical devices. With the development of standard methods for the intentional genetic alteration of the phenotype of research animals, short-lived animals such as laboratory mice often have been used as *in vivo* "laboratories" in which the presence, absence, or altered regulation of genes thought to be related to both normal and abnormal biological processes have been studied (Doyle et al., 2012). The animals in this category are often referred to as "knock-outs" (the elimination of a gene or its function by deletion of all or part of the coding sequence), "knock-downs" (expression of a gene is attenuated by various means), and "knock-ins" (the insertion of sequence information not found within the genome of the animal). The production of such animals has been improbable until highly precise genome targeting was developed with the aid of sequence-specific nucleoproteins, most recently the family of complexes based on CRISPR and associated nucleases (see the previous discussion on genome editing).

With respect to studies intended to benefit animals directly, and humans indirectly, research has included veterinary research to improve the health of animals (primarily companion animals but including larger species such as horses and camels in some countries), or to improve the characteristics of animals intended for animal or human food. With the introduction of techniques for genetic alterations of animal phenotypes, such studies have included the development of animals with resistance to debilitating viral diseases such as porcine reproductive and respiratory syndrome using genome editing (*e.g.*, Proudfoot et al., 2019) and African swine fever (*e.g.*, Hübner et al., 2018). Such studies are often grouped together under the heading of "One Health"[10] which holds as its basic premise the inter-connectedness of human and other animal health, a particularly apposite goal given the recent emergences of zoonotic diseases with pandemic implications.

With the common adoption of genetic alteration in animals as a method of research, academic researchers now generate millions of genome-edited mice every year, and an industry has grown up around contract services to produce genome-edited mice with knock-outs, knock-downs, and knock-ins. Although mice can be useful discovery models with respect to many basic mechanisms in biology, mice and humans diverged from their last common ancestor 65 million years ago and subsequent genetic changes have affected nearly every physiological system. As such, the mouse may not be the ideal animal research model when direct extrapolation to humans is desired. For example, the immune system not only is responsible for eliminating infectious microbes and preventing reinfections but also can generate allergic responses, and dysregulation of the immune system can lead to development of chronic inflammatory diseases (*e.g.*, coronary artery disease, arthritis, diabetes) or autoimmune diseases (*e.g.*, rheumatoid arthritis, lupus, multiple sclerosis). The mouse immune system and the human immune system exhibit numerous differences such as (1) differences with respect to rapid-but-nonspecific immune responses; (2) slower-developing, specific immune responses that involve phagocytosis and killing of microbes; (3) specific immune responses that involve the killing of virus-infected cells; and (4) specific immune responses that involve production of antibodies (Mestas and Hughes, 2004). The immune systems of non-rodent species, including species other than NHP, resemble the human immune system much more closely (Dawson, 2012).

As a result, research animals other than mice are now becoming more widely used in the area of genetically altered animal models. The increased facility with precise genome editing, as well as the ability to propagate animals from a single somatic cell (cloning, or somatic cell nuclear transfer) has led to the ability to produce larger, non-rodent animal models of disease. Pigs are often chosen not only because of their allometric similarities to humans but also because of their similarities to humans in terms of anatomy, physiology, and pathophysiology (Swindle et al., 2012). In a recent review of the use of genetically modified swine (Perleberg et al., 2018), the use of pig models of human diseases, including as models of cardiovascular diseases, cancers, diabetes mellitus, Alzheimer's disease, cystic fibrosis, and Duchenne muscular dystrophy, was described. Genetically altered pigs also have been used as models for human nutritional physiology, and both infant and adult nutrition (Roura et al., 2016).

3.5 GOVERNANCE VERSUS REGULATION

In addition to the complexities of nomenclature regarding the advanced breeding techniques associated with animals and what they are subsequently called, nuances also exist around terms such as "regulation" and "governance". For the purposes of this chapter, "regulation" will be considered the statute-driven authority of a government or governing body to exercise oversight over a product or technology. Statutes are generally considered *as "acts of legislatures that declare, proscribe, or command something; a specific law, expressed in writing"*. Therefore, a statute is a written law passed by a legislature, state, federal, or larger geopolitical unit (*e.g.*, the EU) that sets forth general propositions of law that courts then apply to specific situations.[11] Regulations have been defined as rules of order having the force of law, prescribed by a superior or competent authority, relating to the actions of those under the authority's control.[12]

In the US, most regulatory agencies operate under laws that require them to make "science-based" decisions. These have been described in detail in the NAS[13] report *Animal Biotechnology: Science-Based Concerns* (NAS, 2002), and generally encompass impacts on human, animal, and environmental health. It is important to note that although the US regulation is science-based, the risk analyses that lead to regulatory decisions contain value-based judgements that, in the face of scientific uncertainty, are intended to be protective of human health, animal health, and the environment. Other geopolitical units often expressly incorporate values-based concerns into their laws and regulations (see Section 3.7).

Governance, on the other hand, is a more inclusive, and yet elusive, concept that extends beyond statute-driven regulations. The latter may be limited to science-based concerns, while the former often expands to encompass social responses to scientific concerns, as well as issues dealing with transparency, participation, economics, ethics, equity, among others. In some geopolitical units, such concerns are expressly incorporated into statutes or regulation, so the distinctions may become blurred. We have previously referred to these as "values-based" concerns (Rudenko et al., 2018).

Cvek et al. (2017) have provided an extensive discussion of governance applied to research animals, including the role of Institutional Animal Care and Use Committees (IACUCs) and other mechanisms. The inclusion or exclusion of values-based concerns can result in some tension when attempting to harmonize regulations that are more science-based because "*methods to incorporate social and cultural values into risk analysis are limited because they often cannot be put on the same scale as health risks, environmental externalities, and monetized costs and benefits*" (NASEM, 2017).

3.6 THE SPECIAL CASE OF NON-HUMAN PRIMATES

Discussions of the use of NHP in research tend to be influenced by the competing issues of the closely related genetic and physiological characteristics of NHP and humans, and the ethical issues associated with the use of these animals for research. Primates are defined by zoologists as mammals of the order Primates, comprising lemurs, lorises, tarsiers, New World monkeys, Old World monkeys, and apes including humans. They are characterized by nails on the hands and feet, a short snout, and a large brain.[14]

In the past decade there has been a general recognition that the use of NHP in research should be limited to situations in which another type of animal cannot be used without compromising the scientific findings of the study. A specific case rests on the physiological differences between humans and other types of animals that does not allow for extrapolation of the findings. A report on the findings from a workshop entitled "*International Animal Research Regulations: Impact on Neuroscience Research*" (NRC, 2012) discussed the issue of NHP. The rapporteurs pointed out that "*due to their close phylogenetic relatedness to humans non-human primates are a preferred species for studies directed to understanding issues such as fine motor control, high-level cognitive functions, and decision making*". The rapporteurs also pointed to the findings of two other expert panels that have addressed the issue of using NHP in research (MRC, 2006[15]; MRC, 2011[16]), in which the panels concluded that "*there is scientific justification for the carefully regulated use of non-human primates when there is no other way to address clearly defined questions, including those raised by certain neuroscience studies*". Another example is in drug development for diseases such as HIV/AIDS, certain forms of viral hepatitis, and for testing the safety and quality of some vaccines, most recently those against the coronavirus-19 responsible for the recent pandemic. Thus, even though non-human primate use is declining, the species is still considered an important model in some types of research.

Although there is a recognition of the utility of NHP in research for very specific issues, some Western societies are moving towards limiting or even totally abandoning non-human primate models in medical research. In the US, the National Institutes of Health (NIH) held a workshop in February 2020 focused on ways to improve the rigour and reproducibility of research conducted on NHP. Although participants acknowledged that all research using laboratory animals involves additional obligations relating to the treatment and welfare of their subjects, research with NHP is perceived to bear special obligations.

The EU explicitly prohibits the use of great apes (chimpanzees, gorillas, orangutans, and bonobos) in research, with two exceptions: (1) when the research is directed at the preservation of those species, and (2) when research is for life-threatening, debilitating conditions endangering human beings and no other alternative method or species would suffice. Member States are required to demonstrate this need to the European Commission.

The decline in the use of these species in the West (Chatfield and Morton, 2018) has been paralleled by an increase in non-human primate use elsewhere in the world, often by moving these studies off-shore to countries such as China (Cao, 2018). For example, Cyranoski (2016) stated

> Between 2008 and 2011, the number of monkeys used in research in Europe declined by 28%, and some researchers have stopped trying to do such work in the West. Many have since sought refuge for their experiments in China by securing collaborators or setting up their own laboratories there. Some of the Chinese centers are even advertising themselves as primate-research hubs where scientists can fly in to take advantage of the latest tools, such as gene editing and advanced imaging.

Clearly, the ethics surrounding the use of NHP will continue to be a special circumstance with respect to animal welfare concerns, rules that may develop in this next decade (Figure 3.1).

Figure 3.1 Research involving genetically altered animals and cloning is subject to different legislation in different parts of the world. This can constitute a challenge for research projects spanning several countries/regions and needs to be taken into consideration when planning a research project and ensuring that the necessary permits are in place. (Illustration Anna Kornum.)

3.7 COMPENDIUM OF THE LAWS, REGULATIONS, AND STANDARDS PERTAINING TO RESEARCH INVOLVING GENETICALLY MODIFIED ANIMALS

As previously described in Röcklingsberg, Gjerris, and Olsson (2017), although oversight of much of the research involving genetically altered animals or cloning also falls under the purview of institutional or local Institutional Animal Care and Use Committees (IACUCs), there are laws and regulations that apply to the regulation of animals used for research. The following is a discussion of the framework that governs research involving genetically altered animals and cloning in different parts of the world, including countries with both highly developed economies and those with countries with less economic power. Laws, regulations, and standards in place in individual countries or geopolitical units are summarized in Appendix 1. Included in this table are summaries of the laws and regulations that affect genetically altered animals and cloning. It also includes a summary of the animal welfare laws and regulations for animals in general, which are assumed to apply to all animals in research, regardless of the process by which they are produced. Although this chapter does not discuss every country that has implemented laws, regulations, or standards that pertain to genetically altered animals, they are representative of prevailing regulatory policies and encompass all of the continents with the exception of Antarctica.

3.7.1 Europe

3.7.1.1 THE EUROPEAN UNION (EU)

In 2022, the European Union had a political and economic union of 27 Member States (on January 31, 2020, the United Kingdom exited the EU). Among its accomplishments, the EU has removed barriers to the free flow of commerce, creating a single internal market. Doing so required some overarching harmonization of laws and regulations for those matters in which EU members have agreed to act as one. In general, EU Directives are implemented by individual Member States via their own legislative actions. EU Regulations do not require promulgation by Member States and become active and enforceable once issued. The EU has developed specific regulations related to the processes of biotechnology, including regulations and Directives addressing genetically altered animals and animal cloning for human therapeutic and agricultural purposes (Appendix 1). Key to this chapter are the regulation and governance of commerce related to biological systems (*e.g.*, biomedical issues, food, and agriculture), including import and export.

Regulation EC 1829/2003[17] first set forth the concept that genetically altered organisms, including animals, are regulated according to the process by which those organisms are made. The implication is that those organisms, including animals, will be assumed to be different from traditionally bred animals. As such, mandatory premarket safety assessment by the European Food Safety Authority (EFSA) is required for any food produced via genetic modification. Although EU guidance related to use of genetic modification to produce plants that were then sources of human and animal food appeared in the early 2000s, it was not until 2012 that EFSA published the "*Guidance on the risk assessment of food and feed from genetically modified animals and on animal health and welfare*" that expressly states the following:

> The health status of a food and feed producing animal has traditionally been considered as an important indicator of the safety of derived food and feed and, therefore, the most important component in the risk assessment, addressed in this document, is an extensive comparative analysis of the phenotypic characteristics of the GM animal, including health and physiological parameters. The document also addresses the details of the other components of risk assessment: the molecular characterisation, which provides information on the structure and expression of the insert(s) and on the stability of the intended trait(s); the toxicological assessment, which addresses the possible impact of biologically relevant change(s) in the GM animal and/or derived food and feed resulting from the genetic modification; the assessment of potential allergenicity of the novel protein(s), as well as of the whole food derived from the GM animal; and the nutritional assessment to evaluate whether food and feed derived from a GM animal is as nutritious to humans and/or animals as traditionally-bred animals.
>
> *(EFSA 2012)*

In 2013, the EU issued a second guidance document focused on the risks to environmental receptors, that is, those systems or organisms in direct or indirect contact with the genetically modified animal (EFSA, 2013). The impact of genetically modified animals and their production methods is separately assessed on a case-by-case basis. Additionally, post-market monitoring and surveillance is considered necessary to identify any unintentional effects of the genetic modification on the health and welfare of the genetically altered animal that may arise after marketing.

Animal cloning has been considered separately by the EU; to date, there is no regulation in place, although the EU has put forward two proposed Directives related to animal cloning, one in 2010[18] and the second in 2015 (Appendix 1); on September 8, 2015, the European Parliament voted to ban cloning.

> Taking into account the objectives of the Union's common agricultural policy, the results of the scientific assessments of EFSA based on the available studies, the animal welfare requirement provided in Article 13 TFEU and the citizens' concerns, it is appropriate to prohibit the use of cloning in animal production for farming purposes and the placing on the market of animals and products derived from the use of the cloning technique.[19]

This measure passed by a large margin and went beyond the 2010 Directive proposed by the European Commission, which would have implemented a provisional ban on the cloning of just five species (cattle, sheep, pigs, goats, and horses) as well as going beyond the second proposed Directive in 2015 which limited the ban to only the process of producing a clone, not the products derived from the use of cloning. The arguments for the total ban cited to animal welfare concerns claims that only a small percentage of clones survive to term, and that many clones die shortly after birth. Notably, the EU ban does not cover cloning for research or biomedical purposes and does not address cloning of endangered species. Finally, this ban relates only to the internal EU market, not to imports of clones or their offspring to the EU. Member States will now have to address the issue of cloning on a Member State-specific basis.

In order to understand the reasoning behind some of the actions by the EU Member States and their representatives in Parliament, the amendments made in 2009 to the "Treaty on the Functioning of the European Union" (TFEU) are highly relevant. The 2009 Amendments introduced the recognition that animals are "sentient beings". As Article 13 of Title II states:

> In formulating and implementing the Union's agriculture, fisheries, transport, internal market, research and technological development and space policies, the Union and the Member States shall, since animals are sentient beings, pay full regard to the welfare requirements of animals, while respecting the legislative or administrative provisions and customs of the EU countries relating in particular to religious rites, cultural traditions and regional heritage.[20]

The animal welfare laws and regulations in the EU that existed before 2009 presage animal welfare concerns and biotechnological methods in the 2009 Amendments and have often been cited as the most progressive set of rules for the welfare of agricultural animals. A key feature of all EU animal welfare directives is the emphasis placed on what are referred to as "The Five Freedoms" (see also Chapter 4). These explicitly articulate freedom from (1) hunger and thirst; (2) discomfort; (3) pain, injury, and disease by prevention or rapid treatment; (4) fear and distress; and (5) freedom to express normal behaviour.

The 1986 Directive on the protection of animals used for experimental and other scientific purposes (Directive 86/609/EEC[21]) regulated the use of animals for experimental and other purposes in the EU. It set minimum standards for housing and care, as well as to ensure that individuals performing or supervising such studies had adequate training. This Directive pre-dated the technological development of genetic alterations to living animals. The ultimate goal of this and subsequent Directives on Animal Welfare was to decrease/eliminate the use of live animals, and instead to develop alternative methods (*e.g.*, organ on a chip molecular methods).

In 2010, the EU adopted Directive 2010/63/EU[22] into EU legislation which anchored the principle of the "Three Rs", *i.e.*, to Replace, Reduce and Refine the use of animals in research. It expanded the scope of the previous Directive to include foetuses of mammalian species in the last trimester of development, cephalopods, and animals used for purposes of basic research, higher education, and training. It sought to harmonize, at a high bar, animal welfare standards across the EU. For example, in addition to husbandry and housing standards, it defined a regulated "procedure" to be

> any use, invasive or non-invasive....which may cause the animal a level of pain, suffering, distress or lasting harm equivalent to, or higher than, that caused by the introduction of a needle in accordance with good veterinary practice...including [those] to result in the birth or hatching of an animal or the creation and maintenance of a genetically modified animal line in any such condition....

The 2010 Directive expressly stated its aim to be "*the replacement of procedures on live animals…as soon as it is scientifically possible to do so?...[and] to ensure a high level of protection for animals that still need to be used in procedures*".

A subsequent Regulation added provisions to improve transparency and reporting obligations (EU 2019/1010).[23]

3.7.1.2 GREAT BRITAIN (GB)/UNITED KINGDOM (UK)

Given the recent exit of Great Britain from the EU, changes may be forthcoming in terms of the regulations and standards applying to research using genetically altered animals and cloning. Scientists at the Roslin Institute in Scotland[24] have covered the gamut from cloning Dolly to using genome editing to impart disease resistance in pigs (Lillico et al., 2016). Other key work in the UK is being conducted at Imperial College in London where Target Malaria, an international research programme, is attempting to eradicate malaria. Importantly, the Target Malaria project has an active ethics advisory committee to ensure the "real-time" participation of civil society in areas in which field trials and potential field releases may occur.[25]

Until December 31, 2020, the UK followed EU rules, including those related to development and commercialization of products of various modern molecular technologies such as genome editing, gene drives, and others. Notably, in September of 2021, the UK Department for Environment, Food and Rural Affairs modified its regulatory stance on the requirements for risk assessments of field trials of genome-edited plants, but as of the date of this publication, no such changes have been made for animals. Some UK-specific rules and standards addressing animals in research such as the UK's regulations and standards for animal welfare in research are not affected by the exit from the EU. As a result, there is no separate entry for the UK in Appendix 1 of this chapter.

In terms of animal welfare regulations and GM animals, the use of animals in scientific procedures in the UK is regulated by the Animals (Scientific Procedures) Act of 1986 (amended in 2006). This law reprises, and in some cases exceeds, provisions of European Union Directive 86/609/EEC (regarding the protection of animals used for experimental and other scientific purposes). Relative to other countries, the UK has been considered to afford a relatively high level of protection to animals in medical research, which the law recognizes as necessary. In particular, the UK has additional provisions for the welfare of cats, dogs, and rabbits. Further, the Act's implicit flexibility has allowed the latest technology to be taken into account when deciding whether the use of animals is justified.

Two new pieces of legislation addressing GMOs recently passed into law in England: these are "The Genetically Modified Organisms (Deliberate Release) Amendment" (England Regulations 2019) and "The Animal Health and Genetically Modified Organisms Amendment" (EU Exit Regulations 2019). Recent news reports indicate that the regulations are largely the same as current EU Regulations with one notable exception: the law mandates that GMO regulations are reviewed and revised every 5 years (first report due September 2024) and that the reviews should prioritize, when possible, enacting less "onerous" regulatory provisions.[26] In addition, other changes may be forthcoming in 2023 as the UK is moving to develop new rules around genome editing of animals.

3.7.2 Israel

Although Israel has regulations in place specific to genetically altered plants, there is no such regulation of genetically altered animals. The Ministry of Agriculture's Veterinary Branch is the body responsible for genetically altered animal production experimentation and regulation. This agency evaluates and approves any requests for research or experiments with these animals. In order to import an animal or gamete into Israel, the exporting country must provide a health certificate, and then the Ministry of Agriculture must grant a licence to use that animal or genetic material. Although there does not appear to be a great deal of interest in the genetic alteration of animals for agricultural purposes in Israel (USDA, 2020[27]), and the USDA has stated that "*Genetically engineered animals are not a topic of concern in Israel and there is no legislation or regulation related to the development, trials, commercial use, imports or exports of GE animals*", several Israeli universities are active in research, including on genetically altered animals for biomedical research (*e.g.*, Blechman et al., 2017; Rosenblum et al., 2020; Segev-Hadar et al., 2021). Because this research is being conducted in such academic settings, there is oversight by non-governmental bodies such as university-based animal care and use committees.

The Israeli Animal Welfare Act of 1994 addresses animal welfare including those used in research, although the definition of animal in the law is not specific except to exclude humans. The Law does not privilege one category of animals over another such that companion animals, farm, wild, and captive animals are all entitled to the same level of care and protection. Much of Israeli law and practice regarding animal welfare, including humane slaughter, is based on religious writings and interpretations, and is beyond the purview of this summary. The law appears to regulate but not restrict animal experimentation.

Israel is a member of the OIE, a non-governmental organization comprising 182 member nations whose mission includes protecting animal health and welfare. A recent OIE report outlines animal welfare standards that representatives of some Middle East countries have agreed upon (OIE, 2014[28]). These include the need to prepare a regional strategy that would engage stakeholders and recognize the cultural, religious, and socio-economic differences influencing animal welfare practices within the region. Their report describes general standards for use of animals in research but is only the first step in the process since many Middle East countries lack animal welfare laws, including laws related to animals used in research.

Israel has strict prohibitions against killing animals by poison, with the exception for the prevention of rabies and other zoonoses. With respect to experimentation with animals, regardless of their status of modification, the Minister of Health from the Israeli National Academy of Sciences appoints a chair who oversees the National Council for Research and Development. That body has the legal authority regarding authorizing animal testing and ensuring compliance with standards in place for animal care and use. The members of that Council are representatives of animal welfare organizations. Similar to the US, all institutions that intend to perform animal experiments require an Animal Experiment Permit, which is granted from their Institutional Animal Care and Use Committee. Unlike the US or the EU, there is no specific mandate for an external member of the public on such a panel, although there are representatives of animal welfare organizations participating on the Council. The Council has the authority to prohibit animal use where a "reasonable alternative" exists. An attending veterinarian is appointed whose role is to supervise all studies, and to report to the Animal Care and Use Committee on the progress of research activities and veterinary conditions. The Council has the authority to revoke permits for experimentation.

3.7.3 North America

North America includes three countries with different political systems and different social and economic drivers, although agriculture is a key industry for all three. Canada and the US address products of biotechnology, such as GE animals, in somewhat similar ways, relying on existing regulatory frameworks instead of drafting new laws. Mexico, however, has passed a GMO law that relates to research and commercialization of GMOs including GM animals.

3.7.3.1 CANADA

In 1993, Canada developed the Federal Framework for Biotechnology. Rather than creating a new set of regulations for biotechnology products, the government decided to address novel products produced through biotechnology under existing regulations that cover traditional product categories (*i.e.*, foods, feed, drugs, *etc.*). As a result, products of biotechnology, including genetically altered or GE animals, are regulated by existing standards such as the Canadian Food & Drugs Act (1999). Health Canada, the entity responsible for maintaining and improving the health of Canadians, including food and nutrition, has assumed authority for assessing the safety of products of biotechnology including foods and drugs. GE animals that are fish are assessed prior to marketing by Environment Canada and Health Canada under the authority of the Canadian Environmental Protection Act. As is the case with many countries, "animal" is defined in regulations by its category of use (*i.e.*, laboratory, farm, pet).

When considering definitional issues in Canada surrounding animal biotechnology, the Canadian Food Inspection Agency (CFIA) notes that animal biotechnology includes, but is not limited to animals that are

- Genetically engineered or modified, meaning genetic material has been added, deleted, silenced, or altered to influence expression of genes and traits;
- Clones derived by nuclear transfer from embryonic and somatic cells;
- Chimeric animals that have received transplanted cells from another animal;
- Interspecies hybrids produced by any methods employing biotechnology; and
- Animals derived by *in vitro* cultivation, such as maturation or manipulation of embryos.

With respect to animal cloning, the Canadian regulatory approach was under intensive policy making in the early 2000s, when the Food Directorate issued an Interim Policy of Foods from Cloned Animals (2003). This Interim Policy stated that food from livestock clones would be regulated under the Novel Foods provisions of the Food and Drug Regulations, and that systems in place in Canada for animals and animal products would apply. As a result, clones derived from nuclear transfer from embryonic and somatic cells, their offspring, and the products derived from clones and their offspring would be subject to the same requirements and regulations as those applicable to GE animals and GE animal products. Guidance on the issues has yet to be issued.[29] At the time of this writing, Canada does not require premarket safety assessments for genome-edited products that do not express a novel trait.

Although both federal and provincial animal welfare laws exist in Canada, a 1998 legal opinion issued by the Canadian Council on Animal Care (CCAC) stated that the federal government cannot legislate to protect animals in science, leaving that authority to the provinces and territories.[30] The CCAC was established in 1969; certification from this body is required as a condition for any institution that receives research funding from federal agencies for animal-based projects. The CCAC operates based on four complementary and mutually supportive programmes: the "Assessment and Certification Program" that involves inspection of facilities every 3 years; the "Guidelines Program"; the "Three Rs Program"; and the "Education, Training, and Communications Program". Despite the ceding of animal welfare protection in research to the territories and provinces, the Canadian Federal Government can address criminal acts and health of animals. Sections 444 to 447 of the Criminal Code of Canada protect animals generally from cruelty, abuse, and neglect. The Health of Animals Act (Government of Canada 1990[31]) and its regulations are focused primarily on Canadian livestock, and the Canadian Food Inspection Agency (CFIA) is responsible for testing, inspection, permit issuing, and quarantine activities for live animals (including research animals) imported to Canada.

At the provincial level, all of the provinces have legislation concerning animal welfare, but only certain provinces have legislation specifically addressing animals used for scientific purposes (see Appendix 1). Provincial bodies generally point to the CCAC federal guidelines as having force in terms of research animal welfare. In 2012, the CCAC guidelines were updated to specifically address the welfare of GE animals. Ontario has a stand-alone law entitled "Animals for Research Law 1990". Saskatchewan passed the Animal Protection Act in 2018 that also applies to all animals, including research animals. Quebec recently passed a law (2021) known as the Animal Welfare and Safety Act that applies to all animals, including research animals; in the Quebec law, they identify animals as being "sentient beings".

3.7.3.2 UNITED STATES OF AMERICA (US)

In 1986, the White House Office of Science and Technology Policy (OSTP) published the US Coordinated Framework for the Regulation of Biotechnology[32] (the Coordinated Framework), laying out a comprehensive federal regulatory policy for ensuring the safety of biotechnology products without impeding innovation. To meet the latter goal, the Coordinated Framework stated that no new laws were required to ensure the safety of the products of biotechnology. An interagency working group concurred with the exception of needing additional regulation for some microbial products. Further, they stated that "*The existing health and safety laws had the advantage that they could provide more immediate regulatory protection and certainty for the industry than possible with the implementation of new legislation*". Moreover, there did not appear to be an alternative, unitary, statutory approach since the very broad spectrum of products obtained with genetic alteration cut across many products and uses regulated by different agencies.[33] The 1992 update to the Coordinated Framework further clarified that relevant regulatory agencies would *concentrate on the characteristics of the product and the environment into which it is being introduced, not the process by which the product is created*[34] (*i.e.*, product-based regulation). The most recent update (2017),[35] released in conjunction with the *National Strategy for Modernizing the Regulatory System for Biotechnology Products* (the Strategy), again reaffirmed the US position regarding regulation of the products of biotechnology. It largely reflected the findings of a 2017 study conducted by the *National Academies of Science, Engineering, and Medicine* (NASEM, 2017) that considered what the future products of biotechnology might look like, whether there are likely to be any novel risks, and how the US regulatory system could evolve to concentrating on products with which they are less familiar (*e.g.*, gene drive organisms), instead of applying the same degree of oversight to products with which the US is very familiar (*e.g.*, variations on insect resistance by *Bacillus thuringiensis* in different crops). In essence, there have been no major changes to US biotechnology policy since the issuance of the first Coordinated Framework.

3.7.3.2.1 Regulation under the Coordinated Framework

The three primary agencies with responsibilities for regulation of biotechnology products are the US Environmental Protection Agency (EPA), the FDA, and the USDA. Each agency has its own statutory authority over products and their intended uses,[36] as well as the supporting regulations to achieve health and environmental protection goals (see various laws and regulations listed in Appendix 1). The Coordinated Framework anticipated that statutory authority and agency expertise may not be directly correlated and made provisions for assigning a "lead agency" based on statutory authority, as well as utilizing expertise from across the US government. In exercising their authority, the agencies are to employ a "rational, scientific evaluation" of products and their intended uses. Values-based concerns are not identified as being part of these evaluations.

3.7.3.2.2 USDA

Within USDA, the Animal and Plant Health Inspection Service (APHIS) is responsible for protecting agriculture from pests and diseases. Under the Animal Health Protection Act (AHPA) and the Plant Protection Act (PPA), USDA regulates products of biotechnology that may pose a risk to agricultural plant and animal health, including the importation of animals or animal products into the US. Under the Virus-Serum-Toxin Act (VSTA), APHIS's Center for Veterinary Biologics has regulatory oversight over products of biotechnology that are included in veterinary biologics, and ensures that veterinary biologics are pure, safe, potent, and effective.

APHIS also administers the Animal Welfare Act (AWA) and its standards and regulations. Relevant provisions are that AWA requires basic standards of care and treatment for "certain animals" bred and sold for pets, used in biomedical research, transported commercially, or exhibited to the public. Requirements include adequate care and treatment for housing, handling, sanitation, nutrition, water, veterinary care, and protection from extreme environmental conditions.

Regulated research facilities include hospitals, academic institutions, diagnostic laboratories, and private firms in the biopharmaceutical space. These facilities are subject to unannounced regular inspections or for concerns from the public. It further specifies that research facilities must establish an Institutional Animal Care and Use Committee. The requirements of the AWA and the Animal Welfare Regulations (Code of Federal Regulations, Title 9) (Animals and Animal Products) have been consolidated into what is referred to as the APHIS "Blue Book".[37] Although the AWA regulates the care and treatment of warm-blooded animals, it excludes those used for food, fibre, or other agricultural purposes, pets owned by private citizens, horses not used for research purposes, birds, and rats of the genus *Rattus* and mice of the genus *Mus* that are bred for use in research. It also excludes cold-blooded animals such as snakes, fish, alligators, etc.

The 1986 Health Research Extension Act (HREA) charges the Director of NIH (an Institute under the auspices of the Department of Health and Human Services, as is the FDA) with establishing guidelines for the care of vertebrate animals in the biomedical and behavioural research, teaching, and testing activities funded by NIH.

In addition to the Federal AWA and regulations, many states and local governments can, and often do, establish stricter measures to ensure animal welfare. For example, the State of California has statutes regarding animal cruelty, abuse, and abandonment. The Commonwealth of Massachusetts prohibits animal abuse, neglect, and cruelty; violation is a felony crime in which conviction can, among other things, result in 5 years imprisonment. Texas has civil and criminal laws that apply to livestock and non-livestock animals.

3.7.3.2.3 FDA

The FDA is responsible for protection and promotion of public health. Its statutory authorities include the Federal Food, Drug, and Cosmetic Act (FFDCA), and the Public Health Service Act (PHS). These laws and their enabling regulations provide the basis by which FDA regulates the safety of foods for humans and animal not under the jurisdiction of the USDA, regardless of how they are produced (product, not process regulation). Relevant to this discussion, FDA also regulates (1) the safety and effectiveness of intentional genetic alterations in animals produced using biotechnology under the new animal drug provisions of FFDCA; (2) the safety and effectiveness of human and animal drugs; and (3) the safety, purity, and potency of human biologics, including drugs and human biologics from plants and animals produced using biotechnology. FDA's Center for Veterinary Medicine (CVM) developed a comprehensive risk assessment of the safety of food from animal clones and their sexually propagated offspring[38] that was subject to independent peer review and public comments, as well as a Guidance for Industry on the use of animal clones for human food and animal feed[39] and a risk management plan.[40] These and the agency risk communication approaches have been published (Rudenko and Matheson, 2007).

Under the FFDCA, FDA exercises regulatory oversight of intentionally altered genomic (IAG) DNA in animals under the new animal drug provisions of the FFDCA because the intentional genomic alterations meet the definition of a "drug", *i.e.*, articles intended for use in the diagnosis, cure, mitigation, treatment, or prevention of disease in man or other animals;" or "articles (other than food) intended to affect the structure or any function of the body of man or other animals.

FFDCA (21 U.S.C. 321 et seq.)

It is important to note that FDA does not consider the resulting animal to be a drug; rather it contains an article that meets the definition of a drug. In general, those animals need mandatory approval to enter interstate commerce, unless the agency decides to exercise enforcement discretion (see the subsequent discussion).

The FDA's position regarding the regulatory oversight of IAGs in animals was first articulated in Guidance for Industry 187 (2009)[41]; the draft revision issued on January 19, 2017,[42] expanding the definition of the regulated article from the recombinant DNA construct introduced into the animal to any IAG made to the animal, including by use of highly sequence-specific molecules used for genome editing. Provisions are made for the oversight of animals in the "investigational" phase (*i.e.*, during which research is conducted) including model organisms (*e.g.*, mice, rats, other highly contained physiologically appropriate models) (21 CFR 500.1.(a)) and in the intended species (21 CFR 500.1.(b)). During the investigational phase and extending through to approval, sponsors (parties responsible for the applications) must develop data to demonstrate that the article being regulated is safe to the animal, safe to humans handling the animals, safe for food consumption for humans or animals, and that the issues relevant to the National Environmental Policy Act (NEPA) have been addressed. Exceptions, referred to as enforcement discretion decisions, are made for those animals covered by GFI 187 that are regulated by other agencies, or non-food animal species kept in highly contained laboratory conditions (*e.g.*, mice and rats). In addition, certain other animals with IGAs considered to be low risk, and distributed as animal models of disease, may qualify for enforcement discretion.[43] Examples of the latter include fluorescent aquarium fish (GloFish) and some porcine models of human disease (*e.g.*, minipigs that mimic hypercholesterolemia or cystic fibrosis in ways that mice cannot). Most recently, because of their low-risk status, FDA's Center for Veterinary Medicine recently issued an "enforcement discretion" finding for genome-edited cattle with "slick" coats.[44]

In general, these guidances and regulations apply to all non-human animals containing the regulated article. Depending on the intended use, however, some insects with entomological claims (*e.g.*, releases intended to reduce the populations of *Aedes aegypti*) are considered biopesticides and are regulated by the EPA. Exceptions would occur if public health claims were to be made (*e.g.*, insects whose guts are inhospitable to *Plasmodium falciparum* to combat malaria), in which case FDA would be responsible.[45]

3.7.3.3 MEXICO

Mexico has a separate law that addresses the use of biotechnology. The Law on Biosecurity of Genetically Modified Organisms (GMO Law of 2005) provides a set of rules and standards related to research concerning the release, commercialization, export, and import of GMOs. Due to the breadth of the GMO definition, the regulations can be applied to all organisms that have been genetically altered. The agencies involved in genetically altered animal oversight are the same as for plants and would include the Secretariat of Agriculture (formerly called SAGARPA but now as SADER), the Secretariat of Environment and Natural Resources (SEMARNAT), and the Secretariat of Health (SALUD). The introduction of genetically altered animals for food or feed use would require an authorization from COFEPRIS, the Federal Committee for Protection from Sanitary Risks (Comisión Federal para la Protección contra Riesgos Sanitarios), while the production of genetically altered animals would require a permit from SADER. Although there is no law that would prohibit animal cloning in Mexico, at this time there does not appear to be any cloning of agriculturally relevant animals or any other animals.

Most of the protections in Mexican federal law address animal health rather than focusing on animal cruelty. General animal welfare laws (NOM-062-ZOO-1999, Technical Specifications for the Production, Care and Use of Laboratory Animals (1999), the Act of Animal Welfare (2002), the Health Specifications for Canine Control Centers (2006), the Mexican Official Norm for Disease Prevention and Control (2007), and the Act of General Health and Its Regulation (2015)) apply to all animals; Mexico's Law on Biosecurity of Genetically Modified Organisms (GMO Law 2005) specifically addresses genetically altered animal health as part of the overall assessment by regulatory authorities. One state in Mexico, the Federal District and of Michoacán de Ocampo, recognizes animals as "sentient", although Mexican federal law has not adopted this standard.

3.7.4 South America

In general, countries in South America with the most developed economies, or those that are most dependent on agricultural exports, have specific laws and regulations related to biotechnology, genetically altered animals, and animal cloning and/or animal welfare. This section focuses on countries actively pursuing research in genetically altered animals or animal cloning and on countries that have established standards or guidelines for the use of the technology in animals. Of the three countries discussed below, none have animal welfare laws directed specifically to genetically altered animals or animal clones, although the regulatory authorization process for conducting research in Argentina and Brazil involves consideration of animal health. Chile, in contrast, has animal welfare laws that are more similar to those of the EU because of adopting "sentient being" definitions.

3.7.4.1 ARGENTINA

Argentina was one of the first countries to develop a specific process for regulating genetic alteration methods for plants and animals. In 1991, Ruling Decree 2183/91 was passed by the Argentine legislature and authority was granted to the Ministry of Agriculture, Livestock and Fisheries (MALF) as the first level of oversight. Three other bodies are subject to MALF oversight: (1) the Directorate of Biotechnology (DB); (2) the National Advisory Commission on Agriculture

Biotechnology known as CONABIA; and (3) the National Service of Agrifood Health and Quality, known as SENASA. The regulation of genetically altered animals was put in place to ensure that GMOs are safe for the environment and for human or animal consumption as well as providing a benefit to farmers in Argentina. The rules affect domestic production, export, and import. Regulations require premarket authorization to be obtained from the Secretary of MALF. A recent update to Argentinian law in 2017 (Resolution 79-E) extended the jurisdiction of the government to oversight of both GM animals and animal cloning.

In 2017, the Ministry of Agribusiness, in cooperation with Maimonides University, established the Center for Animal Reproduction and Biotechnology. This new entity is focused on research into development of genetically altered animals for uses that have included the production of pharmaceutical products. Argentina passed a resolution (173/2015) to clarify its position with respect to genome editing. If the technology does not result in the use of a transgene or uses a transgene that is removed in the final product, it is not considered to be a GMO. If a transgene remains in the final animal, it is considered to be genetically modified.

Although animal welfare laws protecting all animals, including research animals (Appendix 1), have existed in Argentina since 1954, no new laws focus on issues related to GM animals or cloning of animals as special circumstances for animal welfare concerns, leaving the oversight within the purview of MALF as part of its authorization process.

3.7.4.2 BRAZIL

Brazil is a leading world producer of GM crops and has had laws and regulations in place since 2005 addressing plant products (Appendix 1). The regulation of genetically altered animals and their products falls under these same laws as modified in 2006 and 2007. The Brazilian Agricultural Research Corporation known as EMBRAPA is a state-owned research corporation affiliated with the Brazilian Ministry of Agriculture involved in biotechnology research in plants and animals. Regulatory oversight of GMOs resides in the National Technical Biosafety Committee (CTNBio) which was formed in 2005 with passage of Law 11.105 (Table GEC). The CTNBio is composed of government representatives, specialists with scientific and technical knowledge in areas of animals, plants, the environment and health, and individuals representing consumers and farmers. Imports of any plant or animal derived from biotechnology, or a by-product that undergoes further processing or is ready-to-consume, must be pre-approved by CTNBio. Reviews for approvals are done on a case-by-case basis. The Bill would place authority for cloning under the Ministry of Agriculture, Livestock and Food Supply (MAPA). It would involve authorization for two different types of animals by two different groups: (1) animals for the production of pharmaceuticals or therapeutics would require authorization under ANVISA, Ministry of Health; and (2) animal clones would require authorization from CTNBio, under the Ministry of Science and Technology. Although review of applications for authorization would involve consideration of animal health, it was not until 2008 that Brazil enacted a law that specifically protected research animals (Appendix 1).

EMBRAPA research has successfully developed genetically altered cattle and goats that produce recombinant proteins. In terms of cloning research, Brazil has a well-developed research system for animal clones under the coordination of EMBRAPA. Cloning research began in the 1990s and focused on cattle. GM animals and animal clones have been developed for commercialization (USDA, 2020a).[46] Commercial Somatic Cell Nuclear Transfer (SCNT) cloning in Brazil is practised by several companies, mostly under a partnership with EMBRAPA. Cloning tends to be focused on cattle for use as elite show or breeding animals. MAPA changed its regulations in 2009 to allow for the genetic registration of cattle clones and their gametes under the Brazilian Zebu Cattle Association (ABCZ); Zebu cattle represents about 90% of the cattle genetics in Brazil. In 2014, CTNBio approved the first commercial release of GM mosquitoes in Brazil, the Oxitec sterile *Aedes aegypti* mosquitoes (OX513A), for field trial purposes, through Brazil's National Health Surveillance Agency (ANVISA). In 2020, Oxitec received full biosafety approval for the *A. aegypti* mosquito from CTNBio.[47]

Because of Brazil's extensive reliance on animal agriculture, animal welfare has been a topic of legislation since 1924 when the federal government issued Decree 16.560 prohibiting behaviours that cause suffering to animals. This was revisited in 1934 when Decree 24.645 prohibited housing animals in conditions that limited breathing, moving, or resting, or deprived animals of light, including abandoning hurt, ill, or "worn out" animals or not providing veterinary assistance, or for rapid euthanasia. Perhaps more importantly, this decree gave animals a standing in federal court where they could be represented by legal counsel. This was illustrated in 2005 by the request for a writ of *habeas corpus* for a chimpanzee housed in allegedly inhuman conditions in a zoo in order to re-home the animals in a Great Apes Sanctuary. Although Judge Edmujndo Lucio da Cruz agreed to hear the case "*in order to incite debate of this issue*", it became moot when the chimpanzee died following two 72-hour requests for extension before a decision was made.[48] In concluding his opinion, Judge da Cruz further wrote "*the topic will not die with this writ, it will certainly continue to remain controversial*".

Article 225 of the Brazilian Constitute of 1988 expressly recognized the fundamental rights for animals such that the government must "*protect the fauna and flora...of all practices which represent a risk to their ecological function, cause extinction of a species, or subject an animal to cruelty*". Perhaps in large part because this Article seems to have applied to wild animals, in 1998, Article 32 of the Environmental Criminal Act criminalized the "*abuse, mistreatment, injury, and mutilation of domestic animals*". Brazil has instituted Good Agricultural Practices that conform with, or exceed, animal welfare standards in other countries, but as Cassuto and Eckhardt (2016) point out, those guidelines are voluntary and, given the importance of animal agriculture to the Brazilian economy, may not become mandatory.

In 2008, Brazil passed Law Number 11,794/2008 which addressed animals in scientific research. It specifically states that animal experiments must be based on ethical considerations and "integrity-based assumptions" (Machado et al., 2009) and provides a regulatory framework by which this could be accomplished. It created the National Council for the Control of Animal Experimentation (CONCEA) to address legal and ethical issues with the use of laboratory animals, relying on the Three Rs. Only institutions accredited by CONCEA can breed or use laboratory animals for teaching or research. CEUA is an ethics committee on the use of animals and is an advisory body formally responsible for the care and use of research and teaching animals within institutions accredited by CONCEA and has the authority to halt any work that is not in compliance with existing legislation. According to Law 11,794, all project involving laboratory animals must be submitted and reviewed by CEUA. Finally, the Law encourages the development of alternative non-animal methods for experimentation (Andersen and Winter, 2019).

3.7.4.3 CHILE

Although Chile is second only to Brazil in South America as an exporter of food, Chile does not have a biotechnology regulatory framework similar to Brazil and Argentina. Chile's first regulation related to genomic alterations was enacted in 1993 by the Ministry of Agriculture's Animal and Livestock Service (SAG for its acronym in Spanish) and focused on the impact of these seeds on agriculture. In 2003, more than a decade after the first seed regulation, genetically altered foods were recognized by an amendment to the National Food Rule which delegated oversight of "biotechnological events" intended for human consumption to the food and drug safety agency in Chile. Although the agency has reviewed petitions for genomic alteration of food events and presumably submitted them for final approval at the Ministry level, after more than a decade, a list of approved such food events has yet to be published (Salazar et al., 2019). With respect to genetically altered animals, there also are no regulations in place to address animals derived from either genomic alteration or cloning. Given the fact that the Animal Protection Act of 2009 established rules regarding animal welfare where animals are identified as "sentient beings", it is not surprising that genetically altered animals and cloning are not being researched in the country.

A recent review of the development of GM products and their use in Chile (Salazar et al., 2019) provides interesting insight into the perception of biotechnology in the country. Chile is described as having a "unique regulatory style" labelled as "the ambivalent regulator", *i.e.*,

> ... when compared with these two [EU and US] approaches, Chile's regulatory style emerges as primarily ambivalent. It shares key aspects of the US familiarity principle policy: no new laws, a product-oriented approach, close adherence to the principle of substantial equivalence and a market regime. Yet it restricts the scale of the technology's use by drawing a boundary between seed-export and all other commercial activity when it comes to defining what is and is not permitted. The restriction on commercial production could be interpreted as a nod to the precautionary principle, which forms the basis of the European stance towards GM crops.

And yet, when Chile's actions are studied, the result has been a vibrant genetically altered seed development and export market but no regulatory progress on the production of such foods, including animals with genomic alterations, for use in Chile.

Animal welfare in Chile is recognized in Law 20380 (the Protection of Animals (2009)), which stipulates that "*animals should be respected and protected as living sentient beings*", and as such has as an objective to avoid the unnecessary suffering of animals. Title IV, Articles 6–10 address the use of animals for scientific research and educational purposes. It further sates that live animal experiments may only be performed by trained personnel, who are trained to avoid animal suffering as much as possible. These articles further specify the requirement of a suitable space for animal housing and surgery, which must be performed by a veterinarian. A permanent Animal Bioethics Committee should consist of two academics, two scientists (authors' note: the distinction between an academic and scientist is not made clear), an investigator, a representative from the Association of Veterinary Medicines, and a representative of institution for the protection of animals with national importance. No specific mention is made as to whether the animals are produced via modern biotechnology. In 2013, an Institutional Committee for the Care and Use of Laboratory Animals was passed in Chile in accordance with a request from OIE.

3.7.5 Asia and South Asia

China, India, Japan, and South Korea comprise this review of the Far East. All four have well-developed economies, and each has a regulatory system in place that addresses GMOs (generally meaning crops) that could generally be applied to GM animals. They have similar broad definitions of what constitutes an "animal" as a living organism that does not include humans. All four countries have animal welfare laws in place that apply specifically to animals used in research; none have specific discussion of animal welfare issues that arise due to use of genomic alterations or cloning. Japan, China, and South Korea are very active in terms of animal cloning, with South Korea focusing on commercial companion animal cloning. Each country recognizes ethical concerns over the use of animals in research and each has developed its own regulatory framework to support the humane care and use of laboratory animals (Ogden et al., 2016).

3.7.5.1 JAPAN

Because Japanese laws refer to "living modified organisms" (LMOs), the same frameworks apply to both plants and animals (Appendix 1). Four ministries are involved: the Ministry of Health, Labor, and Welfare (MHLW); the Ministry of Agriculture, Forestry, and Fisheries (MAFF); the Ministry of the Environment (MOE); and the Ministry of Education, Culture, Sports, Science and Technology (MEXT). In general, for both research and commercialization activities, the approval process involves assessing risks to the environment. Then assessments appropriate to each Ministry are conducted depending on the particular application for the genetically altered animal (*i.e.*, medicine, food, *etc.*). Currently, there appears to be a limited amount of applied research and development for commercialization, with most activity focused on basic research.[49] Animal biotechnology is focused on human medical and pharmaceutical purposes and is performed by universities and government/public research institutions with limited involvement by the private sector. An example of a GM animal that has been commercialized is the GM silkworm for veterinary drug production. It has been suggested that the lack of commercialization of GM animals may be related to negative public reaction towards modern biotechnology, particularly GM and applications to food (Komoto et al., 2016). With respect to cloning of animals in Japan, there are no specific laws or regulations that impose restrictions on animal cloning, where an animal is defined as any non-human organism.

The first national animal welfare law was passed in Japan in 1973 (the Act on Welfare and Management of Animals) and applied broadly to animals of a variety of listed species. The anti-cruelty and duty of care provisions of the law also applied to animals in research (except fish). In 2005, the law was amended to create new basic guidelines for experimentation (research animals) based on the Three Rs (refine, replace, reduce) for animal testing. Despite multiple amendments to the original animal welfare law, key features of many laws in Western countries such as registration of laboratory animal facilities, training of personnel, and regulatory inspections are still not required by Japanese law (Ogden et al., 2016). As a result, self-regulation is part of animal welfare activities in Japan. The laws and regulations specific to GM animals do not discuss any additional animal welfare provisions but such animals would be protected based on the 2005 amendments to the 1973 Act.

With the advent of genome editing and the concomitant discussions regarding whether all of the results of genome editing could or should be considered as producing "GMO" organisms, Japan undertook an analysis to determine, how, as a signatory to the Cartagena Protocol on Biosafety (CPB), they would meet their obligations under the CPB. As previously described, the OECD has characterized three types of editing that can occur; the Japanese government has determined that "*genome edited end products derived by directed mutagenesis... without using a DNA sequence template*" as "*not representing 'living modified organisms'*", SDN2 (template editing) and SDN3 (site-directed insertions) will be considered as LMOs and regulated as such (Tsuda et al., 2019). Hiroshima University has indicated on its website that its research Center for Genome Editing plans to develop "engineered nucleases of Japanese origin" that could be used to develop disease models in cells and animals (frog, mouse, and rat).[50] These would be considered research animals, and it is likely that they would be subject to the same oversight as more conventionally bred research animals. As an example, Tanihara and colleagues (2021) reported on the production of multiple genome-edited pigs for the purposes of producing cells, tissues, or organs for xenotransplantation.

3.7.5.2 CHINA

In 2001, the State Council of China promulgated a general set of regulations related to GM technology and its use in agriculture (Appendix 1). The regulation defined GMOs as "*animals, plants, microorganisms and their products whose genomic structures have been modified by genetic engineering technologies for the use in agricultural production or processing...*" (Chapter 1, Article 3). The regulations provided no further definition of "animal". The Chinese Ministry of

Agriculture and Rural Affairs (MARA) is the regulatory body responsible for approval of biotechnology products for import and domestic cultivation, as well as the development of agricultural biotech policies and regulations.

China has been a leader in GM animal research, launching the Key Scientific and Technological Grant of China for Breeding New Biotech Varieties in 2008. Research has focused on food animal species such as swine, cattle, and sheep, but MARA has not developed a definitive regulatory guideline for GM animals. As a result, China has not yet approved any livestock clones, GM animals, or products derived from GM methods for commercial use. Research has been focused on production of pharmaceuticals, as well as milk and meat quality and quantity.[51] Studies of consumer perception of GE technology in China show that even with the active research efforts, the average person in China has a negative view of GM foods, which would include GM animals used for food production (Cui and Shoemaker, 2018).

Recent news reports indicate that cooperative efforts in China and South Korea have led to ongoing construction of the largest animal cloning facility in the world, a project that reportedly has not been completed.[52] Cloning of both farm animals and domestic pets apparently is to be a focus. This effort may lead to new concerns for animal welfare depending on the focus of the research that develops.

Animal research regulations have been applied to research animals in China since 1988 but the term "animal welfare" did not appear in the regulations until 2006. The definition of "laboratory animals" is broad, applying to any animals that are bred for research purposes and which are used in research of any type. Moreover, a recent article describes the lack of oversight of animal welfare in the research setting in China.[53] As an example of the differences in animal welfare concerns in China versus countries such as the United States, China is the largest supplier of NHP for research (see section on Non-Human Primates in Research).

3.7.5.3 SOUTH KOREA

Laws and regulations specific to biotechnology in South Korea arose from the early desire to promote the tools in science (the 1983 Genetic Engineering Promotion Act, later amended to the 1995 Biotechnology Promotion Act) and evolved to accommodate issues of bioethics[54] particularly after a researcher who had pioneered companion animal cloning was accused and convicted of embezzlement and bioethical violations in research on human cloning.[55] There are very strict bioethics laws governing the cloning of human beings and the derivation of human embryonic stem cells. Cloning animals, especially commercial operations cloning companion dogs and cats, does not appear to be hindered by prohibitions on human cloning.

Laws pertaining to genomically altered organisms and animals were put into place in South Korea in 2006–2008 (Appendix 1). The Living Modified Organisms (LMO) Act does not define an "animal" but uses the broad definition of "living organism" to be "any form of life that can transfer or reproduce genetic material", a definition that would include laboratory animals of all types as well as larger species of food animals. Genetically altered animals that are used for pharmaceutical production are further governed by the Pharmaceuticals Affairs Act, the equivalent to the US FFDCA as it relates to drugs. South Korea is in the process of revising its existing LMO Act (draft released May 2021) to cover products of innovative biotechnologies, including genome-edited products; the draft indicates that there will be a pre-review process that will determine if certain products require a full risk assessment or can be exempt from evaluation.[56]

The Ministry of Agriculture, Food, and Rural Affairs (MAFRA) is the body that is responsible for oversight and approval of genetically altered animals, but no such approvals have occurred. The Ministry of Food and Drug Safety (MFDS) is responsible for safety evaluations for genetically altered animals and fishery products for human consumption and has established a set of safety evaluation guidelines for these animals. Research in the area of genetically altered animals is robust in South Korea as evidenced by the recent creation of a Center to Commercialize New Breeding Technologies and funding of its activities by the Korean government who are developing a transgenic pig model for use in medical research for Alzheimer's therapies (Lee et al., 2017). South Korea also is working in cooperation with the US National Swine Resource and Research Center for further research on transgenic pigs.

With respect to animal welfare laws and regulations, South Korea passed the Animal Protection Act in 2001. It encouraged researchers to use "the least cruelty possible" in animal testing, although it did not describe what that might be. In June 2011, the law was amended to establish a national animal welfare committee and stipulated that animal testing should be approved by that ethics committee before commencing. The Cosmetics Act reform bill (2017) limited animal use in testing, and mandated use of accepted alternative methods, but press reports and non-governmental organizations have reported that the law is not being enforced.[57] In August of 2021, the Ministry of Justice in South Korea made a pre-announcement of legislation to revise the Animal Protection Act to introduce the clause to declare "animals are not things".[58]

3.7.5.4 INDIA

India regulates genetically modified organisms and the products thereof under the "Rules for the manufacture, use, import, export, and storage of hazardous microorganisms, genetically engineered organisms or cells" known as "the Rules". These were notified under the Environment Protection Act (EPA) and are implemented by the Ministry of Environment, Forest and Climate Change, the Department of Biotechnology and State Governments through six competent authorities. The Rules do not specifically address genetically altered animals, but it is likely they are subsumed under the "organism" rubric. In 2021, the Ministry of Health and Family Welfare/Food Safety and Standards Authority of India issued draft regulations for genetically modified organisms or genetically engineered organisms, or living modified organisms intended for direct use as food, which expressly included genome editing under the GMO definition. In April of 2022, the Indian government issued an order exempting SDN1 and SDN2 genome editing products from existing regulation applicable to the products of biotechnology. One of the priority areas listed in the Department of Biotechnology in the

Ministry of Science and Technology is to develop "efficient and high throughput genome editing platforms in a wide range of laboratory animal models" as well as integrating stem cell and genome editing technologies to produce xenotransplantation models.[59]

India has no current regulations on commercial production or marketing of animal clones despite some successes in the cloning of specific species and breeds. Most of the research is done at government research institutions or in academic settings, not by commercial entities. For example, scientists at the National Dairy Research Institute (NDRI) have successfully produced buffalo clones capable of sexual reproduction. At this time the focus of the research is on Indian breeds of buffalo, cattle, and goats (Selokar, 2018).

Before a genetically altered animal can be developed in or imported into India, the Genetic Engineering Appraisal Committee (GEAC) would need to conduct an appraisal of the use. Thus, all aspects of genetic alterations in animal development, including research, are covered by this law. As of 2020, India has not developed any such animals for use in research or for commercial production of foods or therapeutics.[60]

Animal welfare laws define an "animal" as any living creature other than a human. India's earliest legislation related to protection of animals is the PCA Act of 1960 which specifically addressed use of animals in research. Provisions of this Act provided for the establishment of the Committee for the Purpose of Control and Supervision of Experiments on Animals (CPCSEA). Later laws and regulations include the Indian Veterinary Council Act of 1984 and its 1992 amendments as well as the Breeding of and Experiments on Animals Rules that were established in 1998. None of these rules addresses specific issues that would relate to GE or animal clones by themselves but, instead, address any research activity involving animals of all types.

3.7.6 Oceania

The countries included in this overview are Australia, New Zealand, Singapore, and Malaysia. All four have set forth a specific regulatory framework for genetically altered organisms including animals. The framework in Australia is complex because of the existence of federal laws as well as states and territories enacting laws and regulations with different standards. Among the countries reviewed herein, Australia is unique in that it developed animal research and welfare guidelines that explicitly apply to scientific research with genetically altered animals and animal clones. All four countries described in this section have animal welfare laws in place that would extend to animals used in research. New Zealand is unique in the countries considered here because they have defined animals as "sentient beings".

3.7.6.1 AUSTRALIA

The Gene Technology Act of 2000 and its regulations that were put in place also relate to commercialization of genetically altered animals. Under the law, the term "genetically modified organism" or "GMO" includes any kind of organism that has been genetically modified (by modern biotechnology), except humans. The oversight body is the Office of Gene Technology Regulator (EGTR), which then interacts with individual regulatory authorities for different types of products (*i.e.*, Therapeutic Goods Administration or TGA for human medical products; the Food Standards Australia/New Zealand or FSANZ for human food). Because Australia is a Commonwealth, an Inter-Governmental Agreement has been established between the Commonwealth, the states, and the territories, that underpins the system for regulating genetically altered organisms of all types. This agreement resulted in formation of the Legislative and Governance Forum on Gene Technology (LGFGT), a body that comprises ministers from the Commonwealth and each state and territory. The body provides broad oversight of the regulatory framework for such organisms and is involved in policy matters. Support is provided to the LGFGT by the Gene Technology Standing Committee, which comprises senior officials from all jurisdictions. This results in a regulatory process that takes into account the differences that might exist among various states and territories with respect to views on genetic engineering technology.

As part of its GMO laws and regulations, the OGTR maintains a public record of all of its decisions referred to as "dealings". In the 2000 law, research is defined as a "dealing" and all activities or "dealings" involving GMOs are prohibited unless they have been assessed by OGTR to be one of four categories: (1) an Exempt Dealing (these are not listed on the GMO Record); (2) a Notifiable Low-Risk Dealing (NLRD); (3) an Inadvertent Dealing; and (4) an Emergency Dealing Determination. The OGTR has determined that genetically altered animals are considered "Notifiable Low-Risk Dealings" or NLRDs. This is important because a NLRD is defined as "dealings with GMOs that have been assessed as posing low risk to the health and safety of people and the environment provided certain risk management conditions are met". The OGTR maintains a list of NLRDs at their website.[61]

Following public comment, the Gene Technology Amendment Regulation of 2019[62] (modifying the Gene Technology Act of 2000) clarified that if no template was used to direct mutagenesis in an organism (SDN1 techniques), that organism would be considered "exempt" from regulation as a GMO (as would an organism that had experienced that mutation without human intervention). If, however, genome editing used SDN2 or SDN3 techniques (*i.e.*, oligonucleotide-directed mutagenesis, and a template to repair single- or double-strand breaks or involved transgenes) the resulting organisms will continue to be regulated by the OGTR.

In 2007, the National Health and Medical Research Council of Australia developed research and welfare guidelines that explicitly applied to research animals developed with tools of biotechnology.[63] These were amended in 2013 and entitled "Australian Code for the Care and Use of Animals for Scientific Purposes" 8th edition.[64] These guidelines operate under the governing principles of "respect for animals", applying the 3 Rs supporting animal well-being, avoiding harm, pain, and distress. The code has been adopted in all states and territories and provides the principles and ethical framework to guide the decisions and actions of scientists involved in the use of animals for research purposes. With the 2013 revisions to the code, there is now specific reference to, and consideration of, genome altered animals and animal clones used in research. These changes to the Code were informed by the 2012 document produced by the Gene Technology Ethics and

Community Consultative Committee (GTECCC) entitled "National Framework of Ethical Principles in Gene Technology 2012".[65] IN 2014, the NMHRC issued a guide to the "Care of and Use of Australian Native Mammals in Research and Teaching" which incorporates by reference the ethical guidelines in the 8th edition of the code, but provides specific scientific information on native animals, and requires that investigators who have not "previously or recently with a particular taxon [to] seek expert advice from knowledgeable sources before obtaining animals or commencing research".

As shown in the Appendix, genetically altered animals and animal clones also are subject to animal welfare laws or regulations that might exist at the state and territorial level. For example, although there are no federal laws applying solely to animal welfare, the States and Territories regulate animal welfare in their own jurisdictions (Appendix). New South Wales differs from the other jurisdictions in that it has a law specific to research animals (Animal Research Act 1985). All other states and territories incorporate research animal provisions into their general statutes related to animal welfare. Definitions of "animal" also can differ among the states and jurisdictions (Appendix). Two jurisdictions, Queensland and Victoria, also have ethics statements or codes in place that relate to gene technology.[66]

The Commonwealth's Department of Agriculture has addressed animal health issues for animal clones in their import risk assessments. They have determined that animal clones and their products are not considered to be an animal health or biosecurity risk. Like the US, FSANZ does not consider food from animal clones and their offspring to require any additional regulatory oversight because they have determined that food products from clones and their offspring are as safe as food products from conventionally bred animals.

3.7.6.2 NEW ZEALAND

New Zealand has a law and regulations in place that relate specifically to genomically altered organisms: the Hazardous Substances and New Organisms (HSNO) Act (1996) and the Hazardous Substances and New Organisms (Low-Risk Genetic Modification) Regulations (2003). A "GMO" is defined to include "*a plant, animal, insect and micro-organism whose genetic make-up has been changed using modern laboratory techniques*". The HSNO Act regulates both research with such products and materials as well as products released into the market. The Act is administered by the Ministry for the Environment (MfE) and it applies to anything that can potentially grow, reproduce, and be reproduced, including foods and medicines. The Environmental Protection Authority (EPA) administers these regulations. The standards applied by the EPA require that an application for research or commercialization can only be approved if benefits outweigh the perceived risks, with the authority considering effects on the environment, the health and safety of the people, the economy, as well as social and cultural concerns. In 2019, the *Royal Society Te Aparangi* issued a report and series of recommendations on whether existing laws and regulations regarding genetic engineering were sufficient to accommodate genome editing and/or whether certain forms of genome editing would be considered unacceptable for the existing public and cultural views (Royal Society Te Apārangi, 2019). The report recommended additional work that needed to be done to answer those questions. As of October 2021, there have been six applications for contained field trials of genetically altered animals approved by the EPA[67] (USDA, 2020b, 2021).

Unlike genomically altered animals, there are no laws or regulations in New Zealand that apply to animal cloning. This is because clones, by definition, do not have any genetically engineered traits, thus they are not new organisms as defined by the HSNO Act. Any laws that pertain generally to animals, including research animals, would govern the research, use, and management of clones, such as the general animal welfare laws in New Zealand.

In 1999, New Zealand passed into law the Animal Welfare Act of 1999. Animals for use in research are regulated only due to Part 6 of this Act where a research animal ethics committee was established. This meant that research animals were exempt from other parts of the animal welfare law. In 2015, New Zealand amended the 1999 law to recognize all animals as "sentient beings". Additionally, the 2015 amendments specifically addressed genetically altered animals and animal clones and defined "manipulation" in the law to include the need to consider

> breeding or production of an animal using any breeding technique (including genetic modification) that may result in the birth or production of an animal that is more susceptible to, or at greater risk of, pain or distress during its life as a result of the breeding or production.

3.7.6.3 SINGAPORE

The regulatory framework for genetically altered animals in Singapore is the same as the process for plants produced through biotechnology and involves a multi-agency Genetic Modification Advisory Committee (GMAC) established under the country's Ministry of Trade and Industry in 1999. GMAC's objective is to "*ensure public safety while maintaining an environment that is conducive for commercial exploitation of GE products*". The GMAC works in cooperation with the Singapore Food Agency (SFA) and the Ministry of Health (MOH) when making its decisions. The MOH encompasses the Health Sciences Authority (HSA) that has oversight responsibility for pharmaceuticals, medicinal products, and clinical trials, including products made from genetically altered organisms. GMAC may provide advice to the MOH and the HAS, as requested, for genetically altered products, such as those derived using animals. As of 2019, the SFA has replaced the Agri-Food and Veterinary Authority (AVA) as the national body that regulates GM market access in Singapore for products derived from agriculture, which could include GM animals. Like many other countries in Oceania and around the world, animal cloning for research occurs but there are no regulations related to the process.

In Singapore, a genetically altered animal is defined as "*an organism whose genetic material has been altered using genetic engineering techniques*". The GMO guidelines apply to genetically altered animals and are entitled "Guidelines on the Release of Agriculture-Related GMOs" (1999) and "Biosafety Guidelines for Research on GMOs" (2006, revised in 2008 and January 2013). As GMAC is not a regulatory body but an advisory group, their guidelines are not legally binding. The SFA or HAS would need to give final approval on any genetically altered animal application. Under the guidelines, a proposal must be submitted to GMAC for review, after which GMAC decides whether or not to endorse the application. GMAC's decision is then forwarded to the regulatory authority that determines final regulatory approval. GMAC is composed of members from local regulatory agencies and academic institutions, including individuals with relevant expertise. Current genetically altered animal activity in Singapore is limited to research at SFA's Marine Aquaculture Center aimed at development of hatchery technology, including upstream molecular applications, genetic selection to facilitate fish breeding, and the development of fish vaccines and diagnostic kits (USDA, 2019c).[68] Genome editing of animals at the research level is currently ongoing in Singapore, with the Agency for Science, Technology, and Research (A*Star) promoting services for rodents with genome edits.

Animal welfare laws in Singapore date back to 1965 with passage of the Animal and Birds Act and a separate Act called the Wild Animal and Birds Act (1965 as well). For the purposes of animal welfare laws, "animal" is defined as any mammal or fish other than a human being. There is a separate definition for "wild animal" that includes all species of animals and birds of a wild nature excluding dogs, cats, horses, cattle, sheep, goats, domestic pigs, poultry, and ducks. Animals used in research, however, were not covered by these laws. It was not until 2004 that Singapore adopted standards for animal welfare related to use of animals in research (Retnam et al., 2016). These guidelines were established after formation of the National Advisory Committee for Laboratory Animal Research (NACLAR), whose mission was to provide a set of national policies and guidelines for the acquisition, housing, and utilization of laboratory animals in biomedical research. They were developed by adopting best practices from countries such as Australia, Canada, New Zealand, and the United States, and organizations such as the Council for International Organizations of Medical Sciences (CIOMS) and the European Convention for the Protection of Vertebrate Animals Used for Experimental and Other Scientific Purposes (1986). The guidelines are based on scientific, legal, and ethical principles. As a result, research animals of any type, including GE animals and animals that are a product of cloning, would be subject to the NACLAR guidelines.

3.7.6.4 MALAYSIA

The regulatory framework in Malaysia for animal biotechnology is contained in the 2007 Biosafety Act and 2010 Approval Regulations. The Ministry of Natural Resources and Environment, Department of Biosafety, is the regulatory authority for both genetically altered plants and animals. The National Biosafety Board (NBB) is the inter-ministerial body that has responsibility of new genetically altered product reviews. Within the NBB, the Genetic Modification Advisory Committee (GMAC) provides technical, legal and regulatory expertise during the review process. Although this process has been in place since 2010, animal biotechnology product development is not currently ongoing in Malaysia.[69] It should be noted that the Islamic Development Authority of Malaysia (JAKIM) opposes the production and development of animal biotechnology products for the purpose of consumption in Malaysia, which would be expected to impact the development of genetically altered products, including animals. With respect to cloning, like many other countries already discussed, Malaysia has no laws and regulations related to the production of animals by cloning.

Animal welfare laws were first established in Malaysia in 1953 (Animal Act of 1953). The Act defined "animal" broadly such that animal species would be included but it was not until 2013 that the legislation was considered to specifically govern the care and use of laboratory animals and research animals with passage of the Animal (Amendment) Act 2013. Use of non-domesticated species such as NHP and other wildlife in research was addressed by the Wildlife Conservation Act 2010 (WCA 2010). The protection of aquatic mammals and fish is covered under the Malaysian Fisheries Act 1985 (FA 1985). Finally, the Malaysian Animal Welfare Act 2015 (AWA 2015) when considered along with the Animal (Amendment) Act 2013 provide full consideration of welfare of research animals. The other important animal welfare actions in Malaysia relate to adoption of the *Malaysian Code for the Care and Use of Animals for Scientific Purposes* (MyCode) (AWB, 2019).[70] The code is based on the *Australian Code for the Care and Use of Animals for Scientific Purposes 7th Edition* and incorporates elements from the latest Code (the 8th edition). As a result of adoption of the provisions from the Australian Code's 8th edition, language specific to GM animals would apply to animal welfare in Malaysia.

3.8 CONCLUSIONS

In general, the laws, regulations, and other forms of governance related to research animals produced by modern biotechnology (*e.g.*, genome editing, genetic engineering/modification or cloning) continue to develop in geopolitical units around the world. Some of these have specifically addressed the use of modern biotechnology to produce research animals; the European Union, Canada, Mexico, Australia, and New Zealand have amended existing animal welfare laws to accommodate animals produced by modern biotechnology. Others, such as the US and Canada, do not make specific provisions for how the animals are bred, and extend the governance applied to more conventionally bred animals to these animals as well. Geopolitical units also differ in how they define what an animal is, either in general or within the rubric of an applicable statute. Some have only broad categorizations (humans vs all other animals), or whether livestock or companion animals used for research should be regulated differently from those used for other purposes. Regardless, the 3 Rs principle serves as the basis of most research on animal welfare laws, regulations, and standards, regardless of how the animals are produced. In general, the 3Rs principle holds across permissive through precautionary regulatory approaches.

Regulatory systems may also differ in terms of what serves as a "regulatory trigger", *i.e.*, the legal distinction that determines whether, or how, something is regulated. There tend to be two basic approaches: *process*, in which the technology is the trigger for regulation (*e.g.*, cloning or genome editing); and *product*, where the regulatory trigger is often the statutory definition of the regulated article (*e.g.*, in the US, "*an article intended to affect the structure or function of the body of man or other animals*" regardless of how it is produced). It should be noted, however, that despite the differences in regulatory triggers and existing statutory authorities, oversight is conducted on a case-by-case basis. There may be differences in whether such animals may be produced or released depending on the degree of precaution a particular governance system has been implemented.

Sentience is one key difference among the regulatory structures surveyed; the EU, New Zealand, the Canadian Province of Quebec, and Chile each expressly applies this criterion. In the past, sentience has been the basis for imposition of strict ethical rules around human research and even research in some NHP. The use of "sentient" in defining an animal in law in the EU has led to ongoing efforts to replace *in vivo* animal testing with *in vitro* study methods for assessing the safety of products developed by industries such as the pharmaceutical industry and the industrial chemical industry. In the US, where the definition of an animal does not include the term "sentient", similar efforts are underway to limit the use of animals in research overall, indicating an acknowledgement by scientists and regulators that there are questions regarding the ethics of the use of animals in certain types of research. Sentience is also a key consideration when research animals are themselves the target animal being developed (*e.g.*, genome-edited pigs with disease-resistant traits or that serve as large animal models of human diseases). The issue of NHP highlights the concerns regarding sentience in research animals, with conflicting perspectives on whether these animals should ever be used for research, or their use reserved only for those cases in which no other animal model can provide a physiologically relevant system. Some geopolitical units are moving towards banning or extremely limiting use of NHP, while other countries, such as China, appear to be expanding their capacities for research with these animals. This may result in situations similar to "medical tourism" in which research travels from areas in which procedures are not permitted to locations where they are.

Newer biotechnology techniques that are being applied in animal research include production of organisms using genome editing techniques and introducing traits into a population by using gene drives. As with many emerging biotechnologies, statutory regulation tends to lag behind technological development, and the regulatory triggers may involve making distinctions that may not have been encountered in the past. Genome editing as a technique is problematic for process-driven regulation, as the distinction, or regulatory trigger between animals that are considered "GMO", or not, may be a function of the type of genome editing that has been used (*e.g.*, transgenesis vs other mutational outcomes). If all forms of genome editing are considered "GMO", the existing regulatory systems for genetically altered animals can be applied. That is not the case for gene drive organisms which pose another set of regulatory hurdles; although the research and target animals are the same, because the ultimate deployment of these animals may be free release, environmental and ecological concerns are raised in ways that do not necessarily apply to highly contained and confined research animals. To date, few specific statutes or policies advise or specify how to carry out either laboratory studies or field trials using organisms with gene drives, with the exception of the Netherlands, although investigators and funders are developing self-governance systems that comport with existing norms regarding potential risks and benefits (Adelman et al., 2017a, b, Rudenko et al., 2018).

Although this overview is accurate at the time of writing (July 2023), it reflects a snapshot of the current regulatory landscapes across different countries and geopolitical units. Considering the rapid developments ongoing in biotechnology, this summary likely will need updating in relatively short order to accommodate how developing countries will decide to address regulations governing research using modern biotechnology, either as pure research tools or for use in food production, medical therapeutics, or public health.

NOTES

1 Animal and Plant Health Inspection Service (APHIS). 2022. Proposed rule: *Standards for Birds Not Bred for Use in Research under the Animal Welfare Act*. Regulations.gov. https://www.regulations.gov/document/APHIS-2020-0068-8062 (accessed August 7, 2022).

2 The common internationally harmonized definition for "modern biotechnology" is "Modern biotechnology means the application of: (i) In vitro nucleic acid techniques, including recombinant deoxyribonucleic acid (DNA) and direct injection of nucleic acid into cells or organelles; or (ii) fusion of cells beyond the taxonomic family that overcome natural physiological." (FAO, 2003).

3 European Food Safety Authority (EFSA). 'Genetically Modified Animals'. https://www.efsa.europa.eu/en/topics/topic/genetically-modified-animals#efsa-page-title (accessed August 28, 2022).

4 The draft revised guidance addresses animals whose genomes have been intentionally altered using modern molecular technologies, which may include random or targeted DNA sequence changes including nucleotide insertions, substitutions, or deletions, or other technologies that introduce specific changes to the genome of the animal. This guidance applies to the intentionally altered genomic DNA in both the founder animal in which the initial alteration event occurred and the entire subsequent lineage of animals that contains the genomic alteration.
This draft revised guidance has been issued for public comment and has not yet been finalized. Until finalized, final Guidance for Industry 187. Food and Drug Administration. 2015. Guidance for Industry Regulation of Genetically Engineered Animals Containing Heritable Recombinant DNA Constructs https://www.fda.gov/media/135115/download, (accessed March 12, 2023) updated most recently in 2015, remains in effect. https://www.fda.gov/regulatory-information/search-fda-guidance-documents/cvm-gfi-187-regulation-intentionally-altered-genomic-dna-animals (accessed March 12, 2023).

5 U.S. Department of Health and Human Services, Food and Drug Administration (FDA), Center for Food Safety and Applied Nutrition, Center for Veterinary Medicine. 2015. Voluntary Labeling Indicating Whether Foods Have or Have Not Been Derived from Genetically Engineered Plants: Guidance for Industry. https://www.fda.gov/media/120958/download (accessed March 12, 2023).
6 The Food and Agricultural Organization, World Health Organization, Codex Alimentarius. http://www.fao.org/fao-who-codexalimentarius/thematic-areas/biotechnology/en/, (accessed March 12, 2023).
7 Court of Justice of the European Union. 2018. Organisms obtained by mutagenesis are GMOs and are, in principle, subject to the obligations laid down by the GMO Directive. https://curia.europa.eu/jcms/upload/docs/application/pdf/2018-07/cp180111en.pdf (accessed March 12, 2023).
8 U.S. Department of Health and Human Services, Food and Drug Administration, Center for Veterinary Medicine. 2017. Guidance for Industry 187, Regulation of Intentionally Altered Genomic DNA in Animals. https://www.fda.gov/media/133805/download (accessed March 12, 2023).
9 Food and Drug Administration. 2022. FDA Makes Low-Risk Determination for Marketing of Products from Genome-Edited Beef Cattle After Safety Review.https://www.fda.gov/news-events/press-announcements/fda-makes-low-risk-determination-marketing-products-genome-edited-beef-cattle-after-safety-review (accessed March 12, 2023)
10 One Health Initiative. http://www.onehealthinitiative.com/ (accessed March 12, 2023).
11 Legal Dictionary. The Free Dictionary. https://legal-dictionary.thefreedictionary.com/statute (accessed March 12, 2023).
12 Legal Dictionary. The Free Dictionary. https://legal-dictionary.thefreedictionary.com/Government+regulation (accessed March 12, 2023).
13 Prior to 2015, what is now known as the US National Academies of Science, Engineering, and Medicine (NASEM) existed as separate bodies: the National Academy of Sciences (NAS) and National Research Council (NRC), the National Academy of Engineering (NAE), and the Institute of Medicine (IOM). Citations listed in this paper reflect the status of the institutions at the time of publications were issued.
14 The Free Dictionary. https://www.thefreedictionary.com/Primate+(biology) (accessed March 12, 2023).
15 Weatherall, D. 2006. The use of non-human primates in research. [ARCHIVED CONTENT] (nationalarchives.gov.uk) (accessed March 12, 2023).
16 Bateson, P. 2011. Review of Research Using Non-Human Primates. [ARCHIVED CONTENT] (nationalarchives.gov.uk) (accessed March 12, 2023).
17 European Parliament and the Council. 2003. Document 32003R1829. Regulation (EC) No. 1829/2003 of the European Parliament and of the Council of 22 September 2003 on genetically modified food and feed (Text with EEA relevance). https://eur-lex.europa.eu/legal-content/EN/ALL/?uri=CELEX:32003R1829 (accessed March 12, 2023).

18 Official Journal of the European Union, C 311, 12 September 2014. Document C:2014:311:TOC. https://eur-lex.europa.eu/legal-content/EN/TXT/?uri=OJ%3AC%3A2014%3A311%3ATOC
19 European Parliament. 2015. P8_TA(2015)0285 Cloning of animals kept and reproduced for farming purposes.. https://op.europa.eu/en/publication-detail/-/publication/c7e772ff-9f55-11e7-b92d-01aa75ed71a1/language-en/format-PDF (accessed March 12, 2023).
20 European Commission. Animal Welfare. https://ec.europa.eu/food/animals/welfare_en (accessed March 12, 2023).
21 European Council. 1986. Council Directive 86/609/EEC of 24 November 1986 on the approximation of laws, regulations and administrative provisions of the Member States regarding the protection of animals used for experimental and other scientific purposes https://eur-lex.europa.eu/legal-content/EN/ALL/?uri=CELEX%3A31986L0609 (accessed March 12, 2023).
22 European Parliament and the Council. 2010. DIRECTIVE 2010/63/EU OF THE EUROPEAN PARLIAMENT AND OF THE COUNCIL of 22 September 2010 on the protection of animals used for scientific purposes. https://eur-lex.europa.eu/legal-content/EN/TXT/PDF/?uri=CELEX:32010L0063&from=EN (accessed March 12, 2023).
23 European Parliament and the Council. 2019. REGULATION (EU) 2019/1010 OF THE EUROPEAN PARLIAMENT AND OF THE COUNCIL of 5 June 2019 on the alignment of reporting obligations in the field of legislation related to the environment, and amending Regulations (EC) No. 166/2006 and (EU) No. 995/2010 of the European Parliament and of the Council, Directives 2002/49/EC, 2004/35/EC, 2007/2/EC, 2009/147/EC and 2010/63/EU of the European Parliament and of the Council, Council Regulations (EC) No. 338/97 and (EC) No. 2173/2005, and Council Directive 86/278/EE. https://eur-lex.europa.eu/legal-content/EN/TXT/PDF/?uri=CELEX:32019R1010&from=EN (accessed March 12, 2023).
24 Link Roslin Institute Website https://www.ed.ac.uk/roslin/about (accessed March 12, 2023).
25 Target Malaria website. https://targetmalaria.org/ethics-advisory-committee/ (accessed March 12, 2023).
26 Beyond GM. 2019. New UK GMO regulations – what do they mean? https://beyond-gm.org/new-uk-gmo-regulations-what-do-they-mean/ (accessed March 12, 2023).
27 United States Department of Agriculture (USDA), Foreign Agricultural Service. 2020. Agricultural Biotechnology Annual – 2019, Israel's Agricultural Biotechnology Regulations Remain Unchanged https://apps.fas.usda.gov/newgainapi/api/Report/DownloadReportByFileName?fileName=Agricultural%20Biotechnology%20Annual_Tel%20Aviv_Israel_10-20-2019 (accessed March 12, 2023).
28 World Organization for Animal Health (WOAH). 2014. Middle East Regional Animal Welfare Strategy (2014–2019) https://www.oie.int/fileadmin/Home/eng/Animal_Welfare/docs/RAWS_Midlle_East.pdf (accessed March 12, 2023).

29 Food Directorate. 2003. Food Directorate Interim Policy on Foods from Cloned Animals. https://www.canada.ca/en/health-canada/services/food-nutrition/legislation-guidelines/policies/food-directorate-interim-policy-foods-cloned-animals.html (accessed March 12, 2023).
30 Canadian Council of Animal Care (CCAC). https://ccac.ca/en/animals-used-in-science/canadian-legislation-and-policies/ (accessed March 12, 2023).
31 Justice Laws Website. Health of Animals Act (S.C. 1990, c. 21). https://laws.justice.gc.ca/eng/acts/H-3.3/page-2.html (accessed March 12, 2023).
32 Executive Office of the President. Office of Science and Technology Policy. Coordinated Framework for Regulation of Biotechnology, 51 FR 23302, at 23302-23303 (June 26, 1986) https://www.aphis.usda.gov/brs/fedregister/coordinated_framework.pdf (accessed March 12, 2023).
33 1986 Coordinated Framework, 51 FR at 23302-23303
34 Executive Office of the President. Office of Science and Technology Policy. Exercise of Federal Oversight within Scope of Statutory Authority: Planned Introductions of Biotechnology Products into the Environment, 57 FR 6753 (Feb. 27, 1992) (1992 Update to the Coordinated Framework). https://www.whitehouse.gov/sites/default/files/microsites/ostp/57_fed_reg_6753__1992.pdf (accessed March 12, 2023).
35 Obama Whitehouse Archives. 2017. Modernizing the Regulatory System for Biotechnology Products: Final Version of the 2017 Update to the Coordinated Framework for the Regulation of Biotechnology https://obamawhitehouse.archives.gov/sites/default/files/microsites/ostp/2017_coordinated_framework_update.pdf (accessed March 12, 2023).
36 Intended use is an important concept. For example, if FDA were to approve a substance produced by a genetically engineered cow to be used as a drug, that would not mean that the substance would automatically be approved for use in food.
37 United States Department of Agriculture (USDA). 2019. Animal Welfare Act and Animal Welfare Regulations https://www.aphis.usda.gov/animal_welfare/downloads/bluebook-ac-awa.pdf (accessed March 12, 2023).
38 Center for Veterinary Medicine (CVM), U. S. Food and Drug Administration (FDA), Department of Health and Human Services. 2008. Animal Cloning: A Risk Assessment. https://www.fda.gov/media/75280/download (accessed March 12, 2023).
39 Center for Veterinary Medicine, U. S. Food and Drug Administration (FDA). 2008. GUIDANCE DOCUMENT CVM GFI #179 Use of Animal Clones and Clone Progeny for Human Food and Animal Feed https://www.fda.gov/regulatory-information/search-fda-guidance-documents/cvm-gfi-179-use-animal-clones-and-clone-progeny-human-food-and-animal-feed (accessed March 12, 2023).
40 U. S. Food and Drug Administration (FDA). 2008. Animal Cloning, Risk Management Plan for Clones and Their Progeny. https://www.fda.gov/animal-veterinary/animal-cloning/risk-management-plan (accessed March 12, 2023).

41 U. S. Food and Drug Administration (FDA). 2009. Guidance for Industry 187, Regulation of Genetically Engineered Animals Containing Heritable Recombinant DNA Constructs https://wayback.archive-it.org/7993/20170111005939/http://www.fda.gov/downloads/AnimalVeterinary/GuidanceComplianceEnforcement/GuidanceforIndustry/UCM113903.pdf (accessed March 12, 2023).
42 U. S. Food and Drug Administration (FDA). 2017. Guidance for Industry 187, Regulation of Intentionally Altered Genomic DNA in Animals. https://www.fda.gov/files/animal%20&%20veterinary/published/CVM-GFI--187--Regulation-of-Intentionally-Altered-Genomic-DNA-in-Animals.pdf (accessed March 12, 2023).
43 Center for Veterinary Medicine (CVM), U. S. Food and Drug Administration (FDA). https://www.fda.gov/animal-veterinary/animals-intentional-genomic-alterations/intentional-genomic-alterations-animals-enforcement-discretion (accessed March 12, 2023).
44 U. S. Food and Drug Administration (FDA). 2022. FDA Makes Low-Risk Determination for Marketing of Products from Genome-Edited Beef Cattle After Safety Review. https://www.fda.gov/news-events/press-announcements/fda-makes-low-risk-determination-marketing-products-genome-edited-beef-cattle-after-safety-review (accessed March 12, 2023).
45 Center for Veterinary Medicine (CVM), U. S. Food and Drug Administration (FDA). 2017. https://www.fda.gov/regulatory-information/search-fda-guidance-documents/cvm-gfi-236-clarification-fda-and-epa-jurisdiction-over-mosquito-related-products (accessed March 12, 2023).
46 United States Department of Agriculture (USDA), Global Agricultural Information Network (GAIN). 2020a. Agricultural Biotechnology Annual BR2019-0060. https://apps.fas.usda.gov/newgainapi/api/Report/DownloadReportByFileName?fileName=Agricultural%20Biotechnology%20Annual_Brasilia_Brazil_10-20-2019 (accessed August 28, 2022).
47 Link directing to a federal Brazilian website https://www.in.gov.br/web/dou/-/extrato-de-parecer-tecnico-n-7.350/2021-307396425) (accessed March 12, 2023).
48 Legal & Historical Animal Center. https://www.animallaw.info/case/suica-habeas-corpus (accessed March 12, 2023).
49 United States Department of Agriculture (USDA), Global Agricultural Information Network (GAIN). 2020. Agricultural Biotechnology Annual JA2019-0219.
https://apps.fas.usda.gov/newgainapi/api/Report/DownloadReportByFileName?fileName=Agricultural%20Biotechnology%20Annual_Tokyo_Japan_10-20-2019 (accessed March 12, 2023).
50 Research Center for Genome Editing, Hiroshima University. https://www.hiroshima-u.ac.jp/en/ru/aboutcore/Excellence/Genome (accessed March 12, 2023).
51 United States Department of Agriculture (USDA), Global Agricultural Information Network (GAIN). 2020. Agricultural Biotechnology Annual CH2019-0202.

https://apps.fas.usda.gov/newgainapi/api/Report/DownloadReportByFileName?fileName=Agricultural%20Biotechnology%20Annual_Beijing_China%20-%20Peoples%20Republic%20of_10-20-2019 (accessed March 12, 2023).

52 Kraft, A. "China building world's biggest animal cloning factory". CBS NEWS. November 30, 2015. https://www.cbsnews.com/news/china-building-worlds-biggest-animal-cloning-factory/ (accessed March 12, 2023).

53 Rogers, O. "Animal Research in China". Faunalytics. November 19, 2019. https://faunalytics.org/animal-research-in-china/ (accessed March 12, 2023).

54 Sciencescope: "Korea targets lab mischief", Science 315, 5814: 923. February 16, 2007. https://science.sciencemag.org/content/315/5814/923.3 (accessed March 12, 2023).

55 Sang-Hun, C. "Korean Scientist's New Project: Rebuild after Cloning Disgrace". New York Times. March 6, 2014. https://www.nytimes.com/2014/03/01/world/asia/scientists-new-project-rebuild-after-cloning-disgrace.html (accessed March 12, 2023).

56 United States Department of Agriculture (USDA), Global Agricultural Information Network (GAIN). 2021. Agricultural Biotechnology Annual KS2021-0025.
https://apps.fas.usda.gov/newgainapi/api/Report/DownloadReportByFileName?fileName=Agricultural%20Biotechnology%20Annual_Seoul_Korea%20-%20Republic%20of_10-20-2021.pdf (accessed August 28, 2022).

57 Lee, S. "South Korea's cosmetic reform bill has not stopped animal testing, says NGO". Chemical Watch. July 17, 2019. https://chemicalwatch.com/80002/south-koreas-cosmetic-reform-bill-has-not-stopped-animal-testing-says-ngo (accessed March 12, 2023).

58 Ji-Hye, S. "Korea to consider animals not 'things'". Korea Harald. August 19, 2021. http://www.koreaherald.com/view.php?ud=20210819000992 (accessed March 12, 2023).

59 Department of Biotechnology, Ministry of Science & Technology https://dbtindia.gov.in/schemes-programmes/research-development/knowledge-generation-discovery-research-new-tools-and-1 (accessed March 15, 2023).

60 United States Department of Agriculture (USDA), Global Agricultural Information Network (GAIN). 2020. Agricultural Biotechnology Annual IN2019-0109. https://apps.fas.usda.gov/newgainapi/api/Report/DownloadReportByFileName?fileName=Agricultural%20Biotechnology%20Annual_New%20Delhi_India_10-20-2019 (accessed August 28, 2022).

61 Office of the Gene Technology Regulator (OGTR), Australian Government. Notifiable Low Risk Dealings (LNRDs). https://www.ogtr.gov.au/what-weve-approved/notifiable-low-risk-dealings-nlrds (accessed August 28, 2022).

62 Australian Government. 2019. EXPLANATORY STATEMENT, Select Legislative Instrument 2019 No. XX, Gene Technology Act 2000, Gene Technology Amendment (2019 Measures No. 1) Regulations 2019. https://www.legislation.gov.au/Details/F2019L00573/Explanatory%20Statement/Text)(accessed August 28, 2022).

63 Office of the Gene Technology Regulator (OGTR), National Health and Medical Research Council, Australian Government. 2007. Guidelines on the generation, breeding, care and use of genetically modified and cloned animals for scientific purposes (2007). https://webarchive.nla.gov.au/awa/20170816215640/https://www.nhmrc.gov.au/guidelines-publications/ea17
64 National Health and Medical Research Council, Australian Research Council, Australian Government. Australian code for the care and use of animals for scientific purposes. 2014. https://www.nhmrc.gov.au/about-us/publications/australian-code-care-and-use-animals-scientific-purposes
65 Gene Technology Ethics and Community Consultative Committee (GTECCC). 2012. National Framework of Ethical Principles in Gene Technology. https://www.ogtr.gov.au/sites/default/files/files/2021-07/national_framework_of_ethical_principles.pdf (accessed May 16, 2022).
66 Department of Innovation and Information Economy. Queensland. 2001. *Code of Ethical Practice in Biotechnology in Queensland.* https://catalogue.nla.gov.au/Record/1541124 (accessed March 12, 2023).
The Public Health Group, Rural & Regional Health & Aged Care Services Division, Department of Human Services, Melbourne. 2006. Victoria *Statement of Ethical Principles for Biotechnology in Victoria.* http://www3.health.vic.gov.au/biotechnology/downloads/statement1794C0.pdf (accessed March 12, 2023).
67 United States Department of Agriculture (USDA), Global Agricultural Information Network (GAIN). 2021. Agricultural Biotechnology Annual NZ2021-0019. https://apps.fas.usda.gov/newgainapi/api/Report/DownloadReportByFileName?fileName=Agricultural%20Biotechnology%20Annual_Wellington_New%20Zealand_10-20-2021.pdf (accessed August 28, 2022).
United States Department of Agriculture (USDA), Global Agricultural Information Network (GAIN). 2020b. Agricultural Biotechnology Annual NZ2019-0024. https://apps.fas.usda.gov/newgainapi/api/Report/DownloadReportByFileName?fileName=Agricultural%20Biotechnology%20Annual_Wellington_New%20Zealand_10-20-2019 (accessed August 28, 2022).
68 United States Department of Agriculture (USDA), Global Agricultural Information Network (GAIN). 2020c. Agricultural Biotechnology Annual SN2019-0005.
https://apps.fas.usda.gov/newgainapi/api/Report/DownloadReportByFileName?fileName=Agricultural%20Biotechnology%20Annual_Singapore_Singapore_10-20-2019 (accessed August 28, 2022).
69 United States Department of Agriculture (USDA), Global Agricultural Information Network (GAIN). 2020d. Agricultural Biotechnology Annual MY2019-0006.
https://apps.fas.usda.gov/newgainapi/api/Report/DownloadReportByFileName?fileName=Agricultural%20Biotechnology%20Annual_Kuala%20Lumpur_Malaysia_10-20-2019 (accessed August 28, 2022).

70 Animal Welfare Board (AWB). 2019. Malaysian Code of Practice for the Care and Use of Animals for Scientific Purposes. Department of Veterinary Services, Malaysia. https://www.dvs.gov.my/dvs/resources/user_1/2020/Kebajikan%20Haiwan/GP01.pdf (accessed August 28, 2022).

REFERENCES

Adelman, Z., Akbari, O., Bauer, J., Bier, E., Bloss, C., Carter, S. R., et al. 2017a. Rules of the road for insect gene drive research and testing. *Nature Biotechnology* 35: 716–718.

Adelman, Z. N., Pledger, D. and K. M. Myles. 2017b. Developing standard operating procedures for gene drive research in disease vector mosquitoes. *Pathogens and Global Health* 111, 8: 436–447.

Andersen, M. L. and L. M. F. Winter. 2019. Animal models in biological and biomedical research – experimental and ethical concerns. *Anais da Academia Brasileira de Ciencias* 91: e20170238. doi: 10.1590/0001-3765201720170238

Bell, J. and M. Nass. 2015. *Plato's Animals: Gadflies, Horses, Swans, and Other Philosophical Beasts (Studies in Continental Thought)*. Bloomington, IN: Indiana University Press.

Blechman, J., Levkowitz, G. and Y. Gothilf. 2017. The not so long history of zebrafish research in Israel. *International Journal of Developmental Biology* 61: 149–157.

Cao, D. 2018. Ethical questions for research ethics: Animal research in China. *Journal of Animal Ethics* 8, 2: 138–149.

Cassuto, D. N. and C. Eckhardt. 2016. Don't Be Cruel (Anymore): A Look at the Animal Cruelty Regimes of the United States and Brazil with A Call for A New Animal Welfare Agency, 43 B.C. Envtl. Aff. L. Rev. 1, http://digitalcommons.pace.edu/lawfaculty/1018/.

Chatfield, K. and D. Morton. 2018. The use of non-human primates in research. In *Ethics Dumping*, ed. D. Schroeder, J. Cook, F. Hirsch, S. Fenet, and V. Muthuswamy, 81–90. SpringerBriefs in Research and Innovation Governance. Cham: Springer.

Cui, K. and S. P. Shoemaker. 2018. Public perception of genetically-modified (GM) food: A nationwide Chinese consumer study. *npj Science of Food* 2: 10. doi:10.1038/s41538-018-0018-4

Cvek, K., Varga, O. E., Schuppli, C., Ormandy, E., Guillén, J. et al. 2017. Regulation and legislation: Overview and background. In *Animal Ethics in Animal Research*, ed. H. Röcklingsberg, M. Gjerris, and I. Olsson, 91–137. Cambridge, UK: Cambridge University Press.

Cyranoski, D. 2016. Monkey kingdom. *Nature* 532: 300–302.

Dawson, H.D. 2012. A comparative assessment of the pig, mouse and human genomes: Structural and functional analysis of genes involved in immunity and inflammation. In *The Minipig in Biomedical Research*, ed. P. A. McAnulty, A. D. Dayan, N. C. Ganderup and K. L. Hastings. 323–342. Boca Raton, FL: CRC Press.

Diogenes, L. 1959 *Lives of Eminent Philosophers.* Cambridge: Harvard University Press.

Doyle, A., McGarry, M. P., Lee, N. A. and J. J. Lee. 2012. The construction of transgenic and gene knock out/knock in mouse models of human disease. *Transgenic Research* 21, 2: 327–349.

European Food Safety Authority (EFSA). 2013. Scientific opinion: Guidance on the environmental risk assessment of genetically modified animals. EFSA panel on genetically modified organisms (GMO). *EFSA Journal* 11, 5: 3200. https://www.efsa.europa.eu/en/efsajournal/pub/3200 (accessed March 12, 2023).

European Food Safety Authority (EFSA) Panels on Genetically Modified Organisms (GMO) and Animal Health and Welfare (AHAW). 2012. Guidance on the risk assessment of food and feed from genetically modified animals and on animal health and welfare aspects. *EFSA Journal* 10, 1: 2501. https://efsa.onlinelibrary.wiley.com/doi/10.2903/j.efsa.2012.2501 (accessed March 12, 2023).

Food and Agriculture Organization of the United Nations (FAO). 2003. *Safety Assessment of Foods Derived from Genetically Modified Animals, Including Fish.* FAO Food and Nutrition Paper 79. https://www.fao.org/3/Y5316E/y5316e04.htm#bm04 (accessed August 28, 2022).

Friedrichs, S., Takasu, Y., Kearns, P., Dagallier, B., Oshima, R., Schofield, J., et al. 2019. OECD Conference Report: An overview of regulatory approaches to genome editing in agriculture. *Biotechnology Research & Innovation* 3: 208–220.

Garas, L. C., Murray, J. D. and E. A. Maga. 2014. Genetically engineered livestock: Ethical use for food and medical models. *Annual Review of Animal Bioscience* 3: 559–75

Hübner, A., Petersen, B., Keil, G. M., Niemann, H, Mettenleiter, T. C. and W. Fuchs. 2018. Efficient inhibition of African swine fever virus replication by CRISPR/Cas9 targeting of the viral p30 gene (CP204L). *Scientific Reports* 8: 1449.

Komoto, K., Okamoto, S., Hamada, M., Obana, N., Samori, M. and T. Imamura. 2016. Japanese consumer perceptions of genetically modified food: Findings from an international comparative study. *Interactive Journal of Medical Research* 5, 3: e23.

Klymiuk, N., Aigner, B., Brem, G. and E. Wolf. 2010. Genetic modification of pigs ad organ donors for xenotransplantation. *Molecular Reproduction & Development* 77, 3: 209–221

Lee, S., Hyun, H., Park, M., Choi, Y., Son, Y, Park, Y., Jeong, S. et al. 2017. Production of transgenic pig as an Alzheimer's disease model using a multi-cistronic vector system. *PLoS One* 12, 6: e0177933.

Lillico, S. G., Proudfoot, C., King, T. J., Tan, W., Zhang, L., Mardjuki, R. et al. 2016. Mammalian interspecies substitution of immune modulatory alleles by genome editing. *Nature Scientific Reports* 6: 21645.

Machado, C. J. S., Tereza, A., Filipecki, P. and M. Teixeira. 2009. Current Brazilian Law on animal experimentation. *Science* 324, 5935: 1643–1644. http://www.fiocruz.br/omsambiental/media/Artigo_Saldanha_Filipecki_Teixeira.pdf (accessed March 12, 2023).

Mesaros, C. 2014. Aristotle and animal mind. *Procedia - Social and Behavioral Sciences* 163: 185–192.
Mestas, J. and C. C. W. Hughes. 2004. Of mice and not men: Differences between mouse and human immunology. *The Journal of Immunology* 172, 5: 2731–2738.
Murray, J. D. and E. A. Maga. 2016. Opinion: A new paradigm for regulating genetically engineered animals that are used as food. *Proceedings of the National Academy of Sciences of the United States of America* 113, 13: 3410–3413.
National Academies of Sciences (NAS). 2002. *Animal Biotechnology: Science-based Concerns*. Washington, DC: The National Academies Press.
National Research Council (NRC). 2012. *International Animal Research Regulations: Impact on Neuroscience Research: Workshop Summary*. Washington, DC: The National Academies Press.
National Academies of Sciences, Engineering, and Medicine (NASEM). 2017. *Preparing for Future Products of Biotechnology*. Washington, DC: The National Academies Press.
Ogden, B. E., William, W. P., Agui, T. and B. H. Lee. 2016. Laboratory animal laws, regulations, guidelines and standards in China Mainland, Japan, and Korea. *ILAR Journal* 57, 3: 301–311.
Office of Laboratory Animal Welfare, 2015. PHS Policy on Humane Care and Use of Laboratory Animals. https://olaw.nih.gov/policies-laws/phs-policy.htm (accessed on August 28, 2022).
Plato. 2008. *Laws, The Project Gutenberg EBook of Laws*. https://www.gutenberg.org/files/1750/1750-h/1750-h.htm (accessed March 12, 2023).
Perleberg, C., Kind, A. and A. Schnieke. 2018. Genetically engineered pigs as models for human disease. *Disease Models & Mechanisms* 11, 1: dmm030783.
Proudfoot, C., Lillico, S. and C. Tait-Burkard. 2019. Genome editing for disease resistance in pigs and chickens. *Animal Frontiers* 9, 3: 6–12.
Retnam, L., Chatikavanij, P., Kunjara, P., Paramastri, Y. A., Goh, Y. M., Hussein, F. N. et al. 2016. Laws, regulations, guidelines and standards for animal care and use for scientific purposes in the countries of Singapore, Thailand, Indonesia, Malaysia, and India. *ILAR Journal* 57, 3: 312–323.
Röcklinsberg, H., Gjerris, M. and I. Olsson. 2017. *Animal Ethics in Animal Research*. Cambridge, UK: Cambridge University Press.
Rosenblum, D., Gutkin, A., Kedmi, R., Ramishetti, S., Veiga, N., Jacobi, A. M. et al. 2020. CRISPR-Cas9 genome editing using targeted lipid nanoparticles for cancer therapy. *Science Advances* 6, 47: eabc9450.
Roura, E., Koopmans, S. J., Lallès, J. P., Le Huerou-Luron, I., de Jager, N., Schuurman, T. et al. 2016. Critical review evaluating the pig as a model for human nutritional physiology. *Nutrition Research Reviews* 29, 1: 60–90.
Royal Society Te Apārangi. 2019. *Gene Editing Legal and Regulatory Implications*. New Zealand. https://www.royalsociety.org.nz/assets/Uploads/Gene-Editing-Legal-and-regulatory-implications-DIGITAL.pdf (accessed August 8, 2022).

Rudenko, L. and J. C. Matheson. 2007. The US FDA and animal cloning: Risk and regulatory approach. *Theriogenology* 67, 1: 198–206.

Rudenko, L., Pamer, M. J. and K. Ove. 2018. Considerations for the governance of gene drive organisms. *Pathogens and Global Health* 112, 4: 162–181.

Salazar, M. P., Valenzuela, D., Tironi, M. and R. A. Gutiérrez. 2019. The ambivalent regulator: The construction of a regulatory style for genetically modified crops in Chile. *Tapuya: Latin American Science, Technology and Society* 2, 1: 199–219.

Secretariat of the Convention on Biological Diversity. 2000. *Cartagena Protocol.* Montreal: Text and annexes. https://bch.cbd.int/protocol/outreach/new%20protocol%20text%202021/cbd%20cartagenaprotocol%202020%20en-f%20web.pdf (accessed August 8, 2022).

Segev-Hadar, A., Slosman, T., Rozen, A., Sherman, A., Cnaani, A. and J. Biran. 2021. Genome editing using the CRISPR-Cas9 system to generate a solid-red germline of Nile tilapia (*Oreochromis niloticus*). *The CRISPR Journal* 4, 4: 583–594.

Selokar, N. L. 2018. Cloning of breeding buffalo bulls in India: Initiatives & challenges. *Indian Journal of Medical Research* 148, 1: 120–124.

Swindle, M. M., Makin, A., Herron, A. J., Clubb, F. J. and K. S. Frazier. 2012. Swine as models in biomedical research and toxicology testing. *Veterinary Pathology* 49, 2: 344–356.

Tanihara, F., Hirata, M., Nguyen, N. T., Sawamoto, O., Kikuchi, T. and T. Otoi. 2021. One-step generation of multiple gene-edited pigs by electroporation of the CRISPR/Cas9 system into zygotes to reduce xenoantigen biosynthesis. *International Journal of Molecular Science* 22, 5: 2249.

Tsuda, M., Watanabe, K. N. and R. Oshawa. 2019. Regulatory status of genome-edited organisms under the Japanese Cartagena Act. *Frontiers in Bioengineering and Biotechnology* 7: 387.

van Eenennaam, A. L. 2018. The importance of a novel product risk-based trigger for gene-editing regulation in food animal species. *The CRISPR Journal* 1, 2: 101–106.

van Eenennaam, A. L., Wells, K. D. and J. D. Murray. 2019. Proposed US regulation of gene-edited food animals is not fit for purpose. *Npj Science of Food* 3: 1–7.

World Health Organization (WHO) and Food and Agriculture Organization (FAO). 2009. *Foods Derived from Biotechnology.* Rome. http://www.fao.org/3/a-a1554e.pdf (accessed August 8, 2022).

4

Welfare and 3Rs

ANNA KORNUM
Independent Researcher

HELENA RÖCKLINSBERG
Swedish University of Agricultural Sciences

MICKEY GJERRIS
University of Copenhagen

DORTE BRATBO SØRENSEN
University of Copenhagen

DOI: 10.1201/9780429428845-4

4.1 INTRODUCTION

Animal welfare can be seen in various perspectives. And often will the way we think of animal welfare guide the way we identify, measure and treat animal welfare problems. Benny the Beagle could be an example of how our focus will determine our solutions to animal welfare issues. Every time Benny's owner leaves the apartment, Benny starts to bark and whine. And he can go on for hours. Video recordings made by a dog behaviour therapist confirm that Benny is indeed suffering from separation anxiety. Benny's owner feels that Benny's problem is due to Benny's fear of being left alone, so the owner consults a vet and buy some anxiolytic medicine for Benny. This way the anxiety and hence the negative emotions that Benny normally experiences when left alone disappears. Benny's owner has a friend, who thinks this is a wrong approach. Benny's welfare problem, the friend says, is due to the fact that Benny is a pack animal, bred for hunting purposes, and the problem arises because Benny cannot express his natural social pack-living behaviours. This should be solved by either taking Benny to a 'doggy day care' or at least find a neighbour who will take care of him during the day. Just giving medications is not solving the problem, the friend says. Benny's owner and the friend approach animal welfare from two different perspectives and hence their solutions to Benny's welfare problem differ.

In experimental animal research, animal welfare is perhaps a more pressing issue than anywhere else, as animals are sometimes deliberately subjected to pain, fear, disease and injury. In 2015, an estimated 192.1 million animals were used globally for various kinds of scientific purposes (Taylor & Alvarez, 2019). This estimate includes animals used in experimental procedures, animals killed for harvesting tissues and animals bred for laboratory use but not used. In addition, it includes wild-type and genetically modified (GM) animals without a harmful genetic mutation that are used to maintain GM strains.

Of particular interest in relation to GM and laboratory animal welfare are first the unknown traits of GM animals. We may risk phenotypes, in which we are not capable of assessing welfare, recognizing the animal's needs and so on. Secondly, the so-called harmful phenotypes can be seen, where animals, whether intentionally or accidentally, are modified in such a way that they experience pain, fear or distress because of their phenotype or in relation to, e.g., procedures. This highlights the need to focus on the particular welfare risks associated with GM.

There are several reasons why individual scientists or institutions, conducting animal experiments, might be concerned with animal welfare and health, including both purely economic and scientific reasons. For example, it is commonly known that diseased or stressed animals will negatively affect the scientific data. Also, decreased survival in, e.g., a breeding facility will obviously have a negative impact on the economy. The concern for animal welfare may also stem from a sense of direct obligation or sense of care to the animals themselves or simply to accommodate the growing public concern for animal welfare.

Assuming that we have moral obligations towards animals in our care, including experimental animals, we need to be able to assess the animals' welfare. Animal welfare is a concept that can be viewed in many ways, and several definitions of animal welfare have been proposed over the years. In the following, we will briefly introduce first three well-established views or paradigms on the nature of animal welfare (Duncan & Fraser, 1997), and second the concepts of 'hybrid views' or applied frameworks. Last, but not least, we will discuss the 3Rs (Replacement, Reduction and Refinement) which are extensively used as animal welfare principles in biomedical research. In this chapter, we will also deal with potential welfare problems in the three recurring cases (Table 4.1), namely the Tyr-knock-out mice, the heterotaxic zebrafish and the Duchenne muscular dystrophy (DMD) rhesus monkey. The discussion on the intrinsic value of life, i.e., whether death in itself is a welfare problem, will not be included in this chapter (see instead Chapter 5).

Finally, we also discuss what future implications the advances in gene editing and related technologies might have on the welfare of genetically modified animals.

4.2 WHAT IS ANIMAL WELFARE?

When we ask students what they think animal welfare is all about they will suggest that it is important that the animals are well, are feeling good (are neither scared or insecure) and that the animals are able to show normal behaviour. So just like the example with Benny the Beagle, the nature of animal welfare can be perceived in various ways. Philosophers have indeed been discussing theories about what constitutes 'a good life' for more than two millennia. In the course of the recent centuries, these discussions have been extended to include animals and animal welfare, gone beyond examples or fables, and matured into serious philosophies as well as matters of concern to citizens throughout the world. The academic and public discourse has given rise to a range of paradigms based on fundamental assumptions and philosophical perceptions of 'the good animal life', which has led scholars to establishing basic values whose presence is perceived as necessary conditions to ensure a high degree of welfare. These discussions on the nature of animal welfare do not require us to define animal welfare like we would define any technical term, but rather it requires an agreement on which basic values contribute to the welfare of an individual (Tannenbaum, 1991; Duncan & Fraser, 1997). Reflections upon the underlying values and theories on what animal welfare is must naturally precede any attempt to measure animal welfare empirically. However, it is not always obvious by studying a definition of animal welfare how to actually assess the level of welfare as these theories and paradigms may include concepts that are difficult to measure such as happiness, naturalness and preferences.

All the paradigms discussed below have developed traditional empirical methodologies for measuring welfare indicators and interpreting these within their respective conceptual frameworks. Increased sophistication and advances in sciences, such as ethology, neurobiology and physiology, have matured the paradigms from early theories of 'hormonal stress', 'instincts' and an 'ideal of harmony with nature' into modern interdisciplinary sciences (Kirchhelle, 2021). However, it is worth remembering that these measures do not represent an external, objective scale but are reflective of one's conception of what constitutes 'the good animal life'.

Table 4.1 Schematic summary of welfare assessment of the three recurring cases

Example/case	Aim and societal relevance	What is done to the animal?	What will the animal experience?	Severity
Complete knock-out of the Tyr gene in mice (Zuo et al., 2017).	This study was designed to investigate the possibility of a complete knock-out of genes in black or black/dilute brown coloured mice, in a single step using CRISPR/Cas9. Knock-out techniques have long been employed and a wide selection of knock-out animals are already available for basic genetic research and for the study of human and animal diseases. The CRISPR technology could potentially become a more precise, efficient and affordable method to create knock-out animals as disease models, but most studies have reported undesired mosaicism.[a]	Knock-out of the *Tyr* gene for pigmentation was performed at the zygote level, using different versions of the CRISPR method (C-CRISPR). Zygotes were harvested after the natural mating of superovulated[c] females. Superovulation in female mice is most commonly induced by repeatedly restraining and injections with fertility drugs, such as gonadotropins. The females were presumably killed shortly after mating when their fertilized embryos (440) were collected, treated with C-CRISPR and implanted into surrogate mothers. Transfer of embryos is usually performed under general anaesthesia, using a needle and micropipette to inject embryos	Mothers, fathers and surrogates: Although the study does not reveal many specific details, it is obvious that mothers and surrogates have been subjected to invasive fertility treatments, such as superovulation, collecting zygotes from mothers and transference of embryos to surrogates. The procedures mentioned above would entail stressful handling, injections (e.g. in the abdominal cavity), surgery and anaesthesia resulting in post-operative discomfort and/or pain. Mating occurred naturally, which presumably adds to the welfare of the animals.	Mild/moderate

(Continued)

Table 4.1 (*Continued*) Schematic summary of welfare assessment of the three recurring cases

Example/case	Aim and societal relevance	What is done to the animal?	What will the animal experience?	Severity
	The authors showed that adding more single guide RNAs (sgRNA)[b] to direct the Cas9 nuclease (3–4) in a 'Cocktail CRISPR' would result in complete knock-out of the *Tyr* (tyrosinase) gene in mice. Disruption of this gene causes albinism by preventing the production of melanin for pigmentation.	into the oviduct or uterus of surrogates. The inseminated surrogates produced 174 edited mice with complete or various degrees of albinism (mosaicism). Two of the C-CRISPR treatments (3–4 sgRNA) resulted in strictly 100% albino mice (81 mice or ~46%), which indicates that a complete knock-out of both copies of the *Tyr* gene had been achieved. Of the remaining mice in the other groups, between 26% and 77% showed mosaicism of pigmentation. The 100% albino knock-out mice had their blood drawn (from tail) for analysis. Five knock-out albino mice were then naturally mated with normal (wild-type) albino mice and showed normal reproductive abilities, each producing two litter of pups.	The knock-out animals (CRISPR animals): The edited embryos showed normal development and mice born with mosaicism or true albinism would presumably experience little, if any, discomfort because of their condition. We do not know whether the animals (all mice included or produced by the experiment) were killed or used for other purposes.	

(*Continued*)

Table 4.1 (*Continued*) Schematic summary of welfare assessment of the three recurring cases

Example/case	Aim and societal relevance	What is done to the animal?	What will the animal experience?	Severity
Genetic editing (deletion of the mmp21 gene) of zebrafish to enhance the understanding of the complex genetic causality involved in human heterotaxy (Perles et al., 2015).	This study was intended to separate the effects of several genes involved in human heterotaxia by using zebrafish as an *in vivo* model. Heterotaxy syndrome is a condition in which the internal visceral organs are abnormally arranged. Individuals with this condition have various degrees of complex birth defects, affecting, e.g., the heart, lungs, liver, spleen, intestines and other organs. Heterotaxia results in a range of potentially debilitating symptoms.	Zebrafish were used to create an *in vivo* model to establish a causal link between mutations in the MMP21 gene and phenotypic expression of heterotaxia. In zebrafish the mmp21 gene is the equivalent of the human MMP21 gene and scientists achieved a CRISPR/Cas9-induced deletion of mmp21 in zebrafish embryos. To edit the embryos of zebrafish, eggs and semen were collected from donors. The eggs and semen might have been 'stripped', meaning that the germ cells	Depending on the method, the donor fish could have been exposed to stressful handling. The CRISPR-edited fish in the study developed cardiac looping disorders, which can result in a myriad of congenital heart malformations. Human patients, with such disorders, will often exhibit cyanosis from reduced oxygenation/hypoxia. It seems likely that zebrafish would also experience periods of discomfort and lethargy due to hypoxia. The zebrafish in the study were able to grow and reproduce, which indicates a certain degree of normal function.	Moderate

(Continued)

Table 4.1 (*Continued*) Schematic summary of welfare assessment of the three recurring cases

Example/case	Aim and societal relevance	What is done to the animal?	What will the animal experience?	Severity
	According to the U.S. National Library of Medicine, the prevalence of heterotaxia syndrome is 1:10,000 and it is thought to account for approximately 3% of congenital heart defects.	were manually removed from donors, and that fertilization occurred by mixing cells artificially. Stripping of semen and eggs involves handling but is often done under anaesthesia. The deletion of the mmp21 gene resulted in cardiac looping defects, i.e., a defective development of the embryonic heart, in 24% (27/113) of the CRISPR-treated animals (F0 generation). Inter-crossing of five pairs of F0 fish resulted in mutant progeny (F1), where 20%–40% exhibited cardiac looping defects.	We do not know whether the animals (all fish included or produced by the experiment) were killed or used for other purposes.	

(Continued)

Table 4.1 *(Continued)* Schematic summary of welfare assessment of the three recurring cases

Example/case	Aim and societal relevance	What is done to the animal?	What will the animal experience?	Severity
Induced muscle dystrophia in rhesus macaques (Chen et al., 2015).	The direct aim of this research was to assess the potential of the CRISPR/Cas9 method to create large animal models for the study and possible cure for a variety of human diseases, in this case Duchenne muscular dystrophia (DMD). DMD is a recessive form of muscular dystrophy linked to the X-chromosome, it affects around 1 in 3,600 boys. Although the CRISPR technology has shown great potential, the majority of gene-edited animals have, so far, exhibited mosaicism, which can interfere with the desired animal model.	Thirty-two female rhesus monkeys aged 5–8 years were selected as egg donors for superovulation. In this case females were injected with fertility drugs (follitropin alfa) for eight consecutive days. Eggs were collected using a laparoscopic procedure and then fertilized with micro-insemination. The paper does not specify how the semen for fertilization was obtained, but it is possible that it was collected by invasive methods, such as electroejaculation.[d] CRISPR-mediated mutations, resembling the ones found in human DMD patients, were induced in 179 rhesus macaque embryos before they were transferred to 59	Mothers, fathers, and surrogates: Invasive fertilization and obstetric procedures would be expected to have negative impacts on welfare. Stillbirths and miscarriages would also be expected to be uncomfortable. A common side-effect in surrogates of genetically engineered animals is difficult births (dystocia). Finally, the separation of mothers and infants must be a significant source of frustration and emotional stress. The animal models (CRISPR animals): Human DMD patients report no severe physical pain in connection with their condition, but only muscle aches that can be alleviated by non-prescription pain medication. However,	Severe

(Continued)

Table 4.1 (*Continued*) Schematic summary of welfare assessment of the three recurring cases

Example/case	Aim and societal relevance	What is done to the animal?	What will the animal experience?	Severity
	The pathogenesis of DMD has been thoroughly studied in mice, but due to species-specific differences, mice do not express human-like symptoms and primate models will presumably shed further light on the pathogenesis and, perhaps eventually, yield a therapy for DMD.	surrogate mothers. The surrogates underwent several examinations including ultrasonography. The treatment resulted in eight miscarriages, four full-term stillbirths and 14 live-born monkeys, and nine of these were mutants. At the time the study was published, all the live-born infants were still living, but all of them were only about 6 months old and none of	one would expect emotional strain and frustration in both monkeys and humans, due to the inability to perform normal behaviours, such as play behaviour and exploration. From an early age monkeys would probably experience increasingly reduced motor function and eventually paralysis and death, unless the animals are euthanized before the disease becomes advanced. Finally, if the live-born monkeys were hand reared, maternal deprivation could also be a welfare issue.	

(*Continued*)

Table 4.1 (*Continued*) Schematic summary of welfare assessment of the three recurring cases

Example/case	Aim and societal relevance	What is done to the animal?	What will the animal experience?	Severity
	All monkeys were housed at the Kunming Biomed International (KBI), a facility accredited by Association for Assessment and Accreditation of Laboratory Animal Care (AAALAC).	them had yet developed the characteristic symptoms of DMD, which would be expected as the disease is age-dependent. However, tissue samples revealed frequent disruption of normal muscle structure.	We do not know what became of all the monkeys (donors, surrogates, infants born with or without deletion of the dystrophin gene). However, a recent study from the same lab (Wang et al., 2018) reports of a series of tests performed on some of the DMD monkeys. They are currently 4 years old, but no details on their welfare or condition were revealed.	

[a] Mosaicism is the presence of two or more populations of cells with different genotypes in the same individual.

[b] Single guide RNA is a short RNA sequence that guides the Cas9 endonuclease, an enzyme isolated from bacteria, to a site of the targeted gene, depending on the RNA sequence, where the DNA is then cleaved, i.e., edited.

[c] Superovulation is the induced release of multiple eggs.

[d] Electroejaculation is a common method of semen collection in non-human primates kept in labs and breeding facilities. The procedure implies electrical stimulation to induce ejaculation, either by a rectal probe or directly on the penis of monkeys or apes. Some techniques required the animal to be anaesthetized while in other cases, monkeys can be trained to sit relatively still in a chair with restraints.

4.2.1 The functioning-based paradigm

Philosophy and the natural and medical sciences all originate from ancient disciplines that have evolved over time. In the medical sciences, the welfare of a patient has always been the prime focus and something that had to be monitored and assessed when caring for a patient. Medical doctors and veterinarians traditionally look for signs from the body and the mind that indicate that something is out of balance and disturbs the welfare of the patient – and medical symptoms will simply be the indicator that the welfare is in jeopardy. Using such an approach animal welfare tends to focus on the health of the animal and whether the individual functions biologically and is in 'balance' – states that can readily be measured using physiological and behavioural parameters.

The functioning-based paradigm in animal welfare thus focuses on the health and functioning of the biological systems of the animal in which welfare issues are conceptualized as a failure for the animal to cope with its environment, indicated by abnormal behaviour, disease, low growth rates, poor reproductive success, stress, etc. (see, e.g., Broom, 1991) – all symptoms that can relatively easily be measured. This accessibility may lead to the conclusion that the function-based paradigm is objective in its approach to animal welfare. However, this may be debatable as the different research approaches and interpretations that scientists use when assessing animal welfare using a function-based paradigm often reflect value-laden presumptions. Such a conceptualization of animal welfare inherently involves a pre-definition and decision on which values are important when deciding whether disease, low growth rates, poor reproductive success, stress, etc., is better or worse for the animals (Fraser et al., 1997). In other words, using, e.g., low growth rate as a welfare estimate requires a previous decision on why a low growth rate is thought to reduce welfare in the first place. Is it due to a mental negative state brought on by hunger or pathogens; is it due to suppression of adaptive functions, such as maternal behaviour? Moreover, a definition of Animal Welfare based on presence or absence of disease and decreased biological functioning also raises the question if a disease or poor reproductive success is a welfare problem if the individual is not aware of the condition.

Consider a genetically modified male mouse who is sterile due to the modified genome, but otherwise perfectly healthy. Does the sterility then constitute an animal welfare problem? People who endorse the functioning-based approach to animal welfare would probably argue that the welfare of genetically modified animals depends on whether the animal functions normally from a physiological as well as a behavioural point of view and the logical conclusion is then that being sterile will result in reduced welfare even though the animal is not suffering subjectively. It is important to emphasize that this approach to animal welfare also includes that the animals must not be stressed, afraid or in pain, as this will ultimately affect the biological function of the animals and homeostasis/ability to cope.

4.2.2 The feelings-based paradigm

The feelings-based paradigm focuses on the subjective experiences of the animals. Welfare is seen as being free from negative mental states or experiences such as distress, pain, fear, hunger and other negative states. In recent decades, researchers have increasingly emphasized the significance of positive experiences, such as comfort, contentment and pleasure, as relevant when evaluating animal welfare (e.g. Duncan, 1993; Boissy et al., 2007; Lawrence et al., 2018). Included in the feelings-based paradigm is also the fulfilment of preferences often measured by offering the animals a series of options for housing or enrichment resources (Dawkins, 1983, 1990; Cooper & Mason, 2000). If an individual obtains a desired stimulus, it will most likely result in a subjective experience such as pleasure. Unlike the functioning-based paradigm, which can in principle be applied without assumptions of animals having conscious awareness, the feelings-based paradigm presupposes that animals are sentient and have subjective mental experiences that we should care about irrespective of their effect on biological function. Understanding animal welfare in this context entails theories of consciousness, motivation and perception by the animals (for discussion, see Dawkins, 2017).

Understanding animal motivation and subjective experiences can be instrumental to improving animal welfare as to best maximize positive emotions and avoid the negative. In experimental environments it has become increasingly accepted to provide animals with predictability or even controllability, in the form of habituation and training, which can reduce the experienced anxiety and discomfort of laboratory animals when exposed to otherwise aversive procedures, thereby improving animal welfare in the scheme of the feelings-based paradigm (Jønholt et al., 2021).

Assessing animal welfare within the feelings-based paradigm is often difficult as we do not have a "happymeter" that can be connected to the animals' head and measure the level of happiness. Animal welfare must be assessed indirectly through behavioural indicators. As with the physiological indicators discussed in the functioning-based paradigm, such measurements are subject to interpretation and depend in part on the underlying concept of animal welfare. Extensive knowledge of animal behaviour makes it possible to identify and distinguish outwards expressions relating to, e.g., fear, distress, frustration and pain, but also behaviours indicating a positive state of mind such as playing or being relaxed. It is of course vital that these behaviours are correctly interpreted and that behavioural tests capture the complexity of animal behaviour.

In animals with genetic modifications there may be an increased risk that some behaviours may be affected by the genome modifications and that the connection between a behaviour and the corresponding emotional state has been damaged. For instance, a test for sociability called 'the three-chambered box' is sometimes used to gauge how mutant mice with autism risk genes responds to an unfamiliar mouse versus an empty space or an object. Mice that spend more time with the novel mouse are then classified as social. Some scholars have argued that such approaches are oversimplistic at best and that experiments and trials are error-prone when interpretation of complex animal behaviour is reduced to a binary reading. This is particularly the case with mutant strains that might have unexpected reactions to standardized behavioural tests such as inactivity or hyperactivity; both behaviours that are not relating to social behaviour but still may affect time spent close to the other mouse (Katnelson, 2018).

4.2.3 The natural-living paradigm

The natural-living paradigm stresses the possibility for the animal to perform species-specific behaviour in an environment to which the animal is biologically adapted. According to this view, each species has its own inherent and partly genetically determined nature (telos), whereby good welfare implies that the animal can live according to this nature (Rollin, 2004). It is worth noting that 'natural' or species-specific behaviours are considered highly relevant in all the paradigms discussed here, but that it is for different reasons. To the feeling-based paradigm, species-specific behaviours are integrated because they are thought to be of particularly significance to the animal's emotional states (Panksepp, 1998). To the functioning-based paradigm they have adaptive functions (grooming, nesting) which ensures the animal's long-term health ultimately affects biological homeostasis and therefore becomes relevant. To the 'natural-living' paradigms species-specific behaviours are emphasized simply because they are 'natural'.

However, defining the concept of 'naturalness' in a way that makes it suitable for scientific and ethical discussion is not straightforward (Yates, 2018). Should it be understood as the best way to safeguard well-being, as threshold for what is acceptable or is it simply a way of avoiding moral 'wrong-doing' because of the lack of human interference?

Furthermore, the natural-living paradigm is difficult to apply to animals in a captive environment, to domesticated or even more so to GM animals. Proponents of this paradigm will either argue that captivity is inherently incompatible with good animal welfare or that the captive environment should be modified to resemble the animal's natural habitat. However, from the perspective of the animal, not all behaviours considered natural are associated with positive experiences and not all natural lives are good lives (Špinka, 2006). In the wild, animals are faced with numerous threats and stressors such as starvation, pathogens, predators, low-ranking status in social groups, etc. Hence, one might argue that only the positive aspects of natural living should be included, ensuring that the behavioural needs are being met through appropriate enrichment (Baumans, 2005). For example, fleeing from a predator is natural behaviour for a wild mouse. That does not mean that we should introduce cats into our mouse facility. Obviously because it would stress the mice immensely, but also because the mouse is not provided with an environment that enables the mouse to show the full register of behaviour, namely running away to hide in an advanced burrow system.

Determining what is natural for GM animals would to many people seem inherently oxymoronic. For domesticated animals, whether in a laboratory or agricultural setting, it has proven difficult for behavioural scientists to agree upon which behaviours or resources should be considered indispensable (Dawkins, 1983; Jensen & Toates, 1993). Often the behavioural needs of a domesticated species are investigated, at least in part, by studying the behaviour of its wild or feral conspecifics in a natural or semi-natural environment in an attempt to gauge their needs (e.g. Stolba & Wood-Gush, 1989; Kornum et al., 2017). But in the context of GM animals, observing wild conspecifics as a reference is not an option because no wild conspecifics exist as the genetics, i.e., telos, underpinning the behaviour might have been significantly altered. The paramount question relating to GM animals is whether it could, in fact, become impossible to determine which environment would provide the various stimuli that the animal needs to have good welfare.

For the reasons outlined above, it would appear that GM is inherently incompatible with the 'natural-living paradigm'. However, when GM is applied in the name of wildlife conservation, it might be acceptable to some proponents of the natural-living paradigm. In certain cases where invasive species are threatening the existence of endemic species in the wild, so-called 'gene drives' could be applied. Genetic manipulations, editing and cloning could be performed to 'restore' or revert the genome of domesticated or modified animals back to their original wild type, or to increase genetic diversity in endangered species (Grunewald, 2019). Currently various wildlife conservation projects are already turning to biotechnology, exploring the potential of cloning and genetic editing to save endangered species or even to bring back species that have long been extinct, such as the Tasmanian tiger (*Thylacinus cynocephalus*) (Goodyer, 2022). Other applications of GM have already produced tangible results, for instance, in 2020 scientists cloned a 30-year-old specimen of the endangered black-footed ferret (*Mustela nigripes*) in order to increase genetic variation in the wild population (Thomasy, 2021). Despite the potential for conservation many proponents of the natural-living paradigm would presumably still reject GM altogether for being inherently at odds with their conception of animal welfare, either because it represents human interference or that it has been based on decades of fundamental research on animals that were unable to lead natural lives.

4.2.4 Integration of the paradigms

Although the paradigms are here presented as distinct, they are, as already implied, to a large degree intertwined, or coextensive, despite differences in the underlying theories. For example, providing an animal with an environment resembling the animals' natural habitat would in many instances satisfy functional, feeling-based and natural-living paradigms. And poor animal welfare would presumably manifest as such in all the paradigms as impaired function, negative emotional states and abnormal behaviour. However, the paradigms differ when it comes to balancing the different welfare concerns when these come into conflict. Should animals be allowed to range outside despite the increased exposure to pathogens, predators and harsh weather conditions? Should we focus on creating semi-natural laboratory settings or should we train animals to 'feel better' in artificial environments? Such questions expose the underlying philosophical discrepancies. However, because all these paradigms have made valid contributions to how we perceive concepts such as 'animal welfare' and the 'good animal life', many applied frameworks contain elements of all three.

4.3 APPLIED FRAMEWORKS FOR ANIMAL WELFARE

Another, more applicable, way to work with animal welfare is by the use of applied frameworks such as the Five Freedoms or, more recently, the Five Domains (Mellor, 2017). Such frameworks are a mixture of several different philosophical theories as well as more practical, usable concepts, hence including all of the above-mentioned paradigms. The Five Freedoms (FAWC, 1993) were originally developed for use in Farm Animals but can be used for all individuals for which one or more persons have taken on the task of care. In the EU, knowledge of the Five Freedoms forms part of the curriculum in the EU Education and Training Framework for the use of live animals for scientific purposes.

The Five Freedoms are:

1. Freedom from hunger and thirst
2. Freedom from discomfort
3. Freedom from pain, injury, and disease
4. Freedom from fear and distress
5. Freedom to express normal behaviour

The first and third of the Five Freedoms deal with hunger and thirst and pain, injury and disease that are all physiological conditions. As such an animal keeper or veterinarian can respond to these conditions precisely provided this person has the necessary skills to identify – using physiological or behavioural measures or observations – them. The animal can, for instance, be weighed if there is a suspicion of undernourishment, and samples can be cultured if there is a suspicion of infection. Clearly such an approach is based on the biological functioning of the animal. 'Freedom from discomfort', on the other hand, seems to be a feelings-based approach and 'Freedom to express normal behaviour' relates to the natural-living paradigm, since it is important that the animal can express its nature. The Five Domains frameworks contain essentially the same five elements (Mellor & Reid, 1994; Mellor, 2017). The Five Domains have, however, more focus on the emotional state of an animal, acknowledging that when a physical or physiological stimulus is available or lacking, there will most likely be a resulting emotion or subjective experience that can affect the welfare of the individual. For example, if a mouse in a standard laboratory cage perceives humans as a threat and is prevented from showing natural behaviour such as escaping into a burrow system this will most likely induce fear and distress in the mouse. Another example could be providing cereals and grains hidden in the bedding. This will allow for natural foraging behaviour and most likely result in a positive mental state when these preferred, palatable foods are found and eaten.

4.3.1 Animal welfare in laboratory animal science: The 3Rs

Traditionally the welfare of laboratory animals has not been discussed using the above-mentioned frameworks or paradigms but in terms of the principles known as the 3Rs (Replacement, Reduction, Refinement). The three principles were introduced in 1959 by Russel and Burch in their book *The Principles of Humane Experimental Technique*, to ensure 'humane' practice as well as to improve the quality of the research. The principles of the 3Rs were classified as ways to diminish or remove inhumanities in *in vivo* research. The 3Rs have profoundly impacted research practices and policies around the world, and are, e.g., implemented in the EU legislation (the Council directive 2010/63/EU article 4). Simply put, these principles state that, if possible, methods which avoid or replace the use of animals should be used (Replacement). Methods which minimize the number of animals used per experiment should be applied (Reduction) and last, methods which minimize suffering and improve animal welfare for the animals still to be used should be implemented (Refinement). For obvious reasons Refinement tends to be the most widely discussed principle for people working directly with the animals.

Refining includes the implementation of all possible methods to reduce inhumanities-inhumanities being conceptualized as distress caused by, e.g., pain, fear, anxiety and conflict- both in the laboratory and housing facilities. However, it is also stated that animal behaviour should be evaluated on a scale or spectrum ranging from what Russell and Burch call "*complete well-being to acute distress*" and they conclude that "*In this way, we are led ... to attempt always to drive the animals up to the highest possible point of the scale. Thus we can aim at well-being rather than at mere absence of distress*" (Russell & Burch, 1959/Chapter 2). Russell and Burch do not offer any definitions on animal welfare or well-being and the 3Rs is therefore an overall, including concept that seems to be in accordance with the above-mentioned Five Freedoms and Five Domains, and also with the three examples of animal welfare paradigms. Basically, the 3Rs state that anything that can be done to reduce pain, fear, anxiety and conflict and enhance well-being in the broadest sense of the word should be done.

4.3.2 Perspectives of the 3Rs, animal welfare and GM

Altering the genetic material of an animal would, in all the paradigms discussed, be viewed as a potential welfare hazard due to the uncertainties involved, for example, pleiotropic effects of the altered genes could negatively impact the animal's biological function or subjective experience, or they might violate the natural state/telos of the animal.

In relation to the 3Rs (primarily Reduction or Refinement) editing technology could, despite the risks, potentially have beneficial effects on animal welfare, as will be discussed below.

4.3.3 Refinement: contingent vs. deliberately induced suffering

In 1959, the debate on GM animals was not yet relevant. However, with the increasing use of, e.g., the CRISPR technique, the possible appearance of unexpected phenotypes leading to distress or suffering in the animals becomes a reality. Hence, the distinction between the possible direct suffering relating to the deliberately genetically modified phenotype and the contingent suffering due to an unexpected, additional phenotypic trait is becoming an important focus point.

Despite the overarching approach to the definition of inhumanities, Russell and Burch make an important distinction between direct and contingent inhumanity:

> We must first distinguish direct and contingent inhumanity. By the former, we mean the infliction of distress as an unavoidable consequence of the procedure employed, as such, even if it is conducted with perfect efficiency and completely freed of operations irrelevant to the object in view. (It does not, of course, follow that a given procedure is the only means of obtaining the desired information, or that it cannot be replaced by a less directly inhumane method.) By contingent inhumanity, on the other hand, we mean the infliction of distress as an incidental and inadvertent by-product of the use of the procedure, which is not necessary for its success. In fact contingent inhumanity is almost always detrimental to the object of the experiment, since it introduces psychosomatic disturbance likely to confuse almost any biological investigation.
>
> *(Russell and Burch, 1959/Chapter 4)*

This distinction is highly relevant in laboratory animal research, where the primary objective is to induce human diseases or injuries, which seldom can be done without inflicting the often harmful conditions on the animals. The crucial point is then to eliminate all contingent suffering, making the unavoidable, direct suffering the only cause of distress in the animals. And in relating to GM animals, Refinement needs to be a continuous focus including all strains produced with GM techniques. It must be ensured that any unintended impacts such as off-target effects and pleiotropic consequences on phenotypic traits do not add to contingent suffering. Due care and diligence including thorough and conscious evaluation of the physiology and behaviour of these animals should be employed to avoid – to the highest possible extent – any contingent suffering.

4.3.4 Could GM help refine and improve animal welfare?

Some have argued that more accurate technologies could be advantageous. Tools such as CRISPR could eliminate unintended pleiotropic effects and thus reduce unintended harms. Many of the 'classic' laboratory animals, e.g., the athymic nude mouse, have been created by selective breeding of animals with naturally occurring mutations. Such artificial selection of mutation-based phenotypes may involve unintended effects. An example would be the NOD-mice (non-obese diabetic mouse) created for the study of diabetes mellitus, but which also accidentally suffer from the autoimmune disease Sjögren's syndrome (SS), mainly affecting salivary and lacrimal glands. However, older animal models are now being replaced by new generations of carefully tailored genetic lines, e.g., the nude immune-deficient mice, often used in cancer research and xenografts, have in many cases been replaced by RAG mice, which are still immune-deficient but, for example, have maintained fur coat cover and therefore may have a better ability to thermoregulate compared to nude mice. New disease models can time the emergence of the disease and thereby reduce the time span of experiments, and genes can in a sense be 'turned on or off' when it is required, sparing the animals a lifetime of symptoms. One might imagine cases where the genetic alterations could even enhance biological functions, such as immunity, resulting in an animal more resistant to infections. Or perhaps a genetic alteration could potentially result in an animal with a reduced stress response or even a more up-beat mental disposition. (Several loci have already been linked to mental dispositions, e.g., variants of the serotonin transporter (5-HTT, SLC6A4) gene promoter or the Oxytocin Receptor Gene (OXTR).) In such cases, we need to ask whether the animal is still being harmed, since we are no longer able to measure it using stress-related biomarkers or fear-related behaviours. It is already possible to genetically modify mice to be less sensitive to pain (Devolder & Eggel, 2019; Remmel, 2021) which could be beneficial for mouse welfare in studies inducing pain as part of the model in, e.g., arthritis studies. In these studies, inflammatory arthritis is induced and since the disease is the phenomenon being studied, pain relief cannot be administered as it would significantly change the course of the disease. Several mutations that result in either complete or partial insensitivity to pain have been known for decades and have been described in human patients who can verbally report on their sensations or lack thereof (Foulkes & Wood, 2008), which would seem ideal for mice being subjected to experimental procedure or might have been born with an otherwise painful condition. Being less sensitive to pain may very well benefit these mice, provided of course that their immune system is not affected by the mutation. On the other hand, it could be speculated that such mice may behave abnormally, in general, and injure themselves in social interactions (e.g. escalating male aggression) or other aspects of their caged lives.

4.3.5 Will GM result in reduction?

Reducing the overall number of animals can be undertaken in several ways by, e.g.

1. reducing research activity, which would be considered undesirable by many as the amount of knowledge gained would presumably also decrease accordingly;
2. replacing animal experiments with alternative methods (see below and Chapter 2), which would be ideal in the framework of the 3Rs as long as the alternative methods were adequate or better than animal experiments;

3. 're-using' animals, which would in many cases be undesirable to the animals being exposed to repeated procedures and furthermore it would in many cases violate the principle of refinement.

It is uncertain whether new, more precise, editing tools will ultimately result in a decrease or an increase in the total number of animals being used for scientific purposes. Although the enhanced precision has already reduced the number of animals needed to create and maintain new strains of genetically altered animals in some situations, ironically it has also unleashed a variety of, previously unavailable, possibilities, which seem to be driving an increase in animals used in other situations (Schultz-Bergin, 2018).

With traditional vector-mediated approach and GM techniques using embryonal stem cells, it would often take several generations to create a modified strain of mice. Creating novel animal models using GM has become increasingly fast and cost-effective. The creation of a new type of transgenic mice has been reduced from 1–2 years on average to 1–2 months using CRISPR. There has been a surge in novel animal models, including exotic animal models such as KO squid and cuttlefish (Reardon, 2019). All of these would be expected to increase the overall number. However, there are also indications to the contrary, e.g., in the European Union, the number of animals used for creation and maintenance of genetically modified lines has decreased by 20% despite a 7% increase in creation of new lines (European Commission, 2020). The global fraction of GM animals in research is rising. Although an estimate of the total number worldwide is not currently available, in the EU around 27% of animals used for the purpose of research were genetically modified in 2017 and 17% of these were considered as exhibiting harmful phenotypes (European Commission, 2020).

4.3.6 Will GM result in replacement?

Like in the case of Reduction it is at this point difficult to project exactly how genetic editing will affect the efforts to replace animals with alternative methods (see also Chapter 2).

4.4 THE THREE CASES

Assessment of welfare in the three cases (for overview, see Table 4.1) of the Tyr gene in mice, heterotaxy in zebrafish and DMD in non-human primates is obviously encumbered by the scarcity of details provided by the scientific papers that reported them as well as knowledge gaps of species and of the induced conditions.

Procedures in laboratory animals subjecting the animals to pain, suffering, distress and/or lasting harm have been categorized as 'mild', 'moderate', 'severe' or 'non-recovery'. Practices that are not likely to cause pain, suffering, distress and/or lasting harm equivalent to or higher than that of an injection done correctly are – at least in the EU – considered to be below this "injection criteria" threshold and therefore such animal experiments do not need to be authorized by a competent authority.

For animal experiments, in general, the potential harm often must be predicted in advance. Researchers must supply information about relevant factors such as husbandry conditions, species of animal, the expected duration and pain, harm and suffering involved in the procedures the animals are to undergo as well as endpoints and killing method. When working with GM animals, adequate information on the harmfulness of a genotype is often only available after a series of generations, which adds to welfare risks, when choosing GM animals as experimental organisms.

- **Knock-out mice:** In the present case the partial or complete knock-out of the Tyr gene in mice would not be expected to compromise welfare except for the general risks associated with albinism in mice, at least not from a functioning- or feeling-based perspectives provided that the animal is reared in a protected environment where its albinism does not make it extra vulnerable to ultraviolet radiation and predators. Many strains of laboratory mice (balb/c and NMRI) and rats (Sprague Dawley and Wistar rats) are albino animals, and this is normally not considered as adding to the cumulative load on the animals. By most accounts, this case would be classified as 'sub-threshold' or 'mild'. However, the fertility treatment of donors and surrogates would be expected to be associated with pain and discomfort and therefore problematic from all animal welfare accounts.
- **Heterotaxy in zebrafish:** When making welfare assessments for teleost fish one is immediately faced with a number of challenges. Compared to most mammalian and avian species the available knowledge of fish physiology and behaviour in general is limited (Huntingford et al., 2006; Balcombe, 2016). The evolutionary distance as well as the alien physical medium they inhabit makes it difficult to apply the usual protocols to fish and some scholars are still questioning their ability to have 'conscious experience' of morally relevant sensations, such as pain (Rose et al., 2012; Key, 2016). However, there are structural similarities between the central nervous system of fish and 'higher' vertebrates (Butler & Hodos 1996/2005; Algers et al., 2009). There is anatomical, pharmacological, and behavioural evidence to suggest that fish are likely to experience affective states of pain, fear and stress (Chandroo et al., 2004; Braithwaite et al., 2013; Braitwaite & Ebbesson, 2014). Fish show behavioural signs of fear when faced with a novel object (Sneddon et al., 2003) or fleeing a predator.

  Zebrafish are an increasingly popular vertebrate model in toxicology, drug-development and the study of human diseases, yet most protocols for the care of zebrafish are primarily concerned with productivity and generally places little emphasis on welfare which is in part reflective of the fundamental knowledge gaps (Lee et al., 2022). In relation to the induced heterotaxia, the harm is difficult to classify for reasons outlined above. However, handling of donor fish, cardiac looping with increasing risk of hypoxia and resulting lethargy would presumably have a negative impact on the welfare of a 'moderate' severity.

- **Induced DMD in rhesus macaques:** The use of non-human primates has for many years, been the subject of controversy in both activist and scientific communities. The close evolutionary relation to humans, the physiological similarities and level of intelligence are often highlighted as justification for non-human primates as ideal animal models but equally to make the opposite case, i.e., why they ought to be given special moral consideration (Sauer, 2000). A variety of animal models for DMD are already available, but all of them differ in relevant ways from human patients. Murine and canine models have had successful introduction of DMD. However, mice have milder phenotypical expressions compared to human DMD patients and dogs born with dystrophin deletion have shown extremely varying severity among littermates (Lim et al., 2020).

Rhesus macaques have for the last few years been used to model a variety of neurological and neurodegenerative diseases, including Huntington's, Alzheimer's and autism and precisely because of their neurological resemblance to humans they are considered particularly well suited as models.

The case of DMD is not straightforward, partly due to the intelligence and cognitive complexity of the macaques. Although Duchenne dystrophy in humans is only associated with moderate dull pain, the behavioural deprivation as well as anxiety associated with poor mobility is presumably severely harmful. Muscle dystrophy might be evaluated significantly differently in the 'functioning-based' and 'feeling-based' paradigms, at least in the early stages until the condition becomes fully expressed. Apart from the induced condition in itself, a range of contingent harms are inflicted on the modified animals as well as the mothers and surrogates involved. Apart from various fertility and insemination procedures, the hand-rearing of the infant macaques is especially problematic as macaques form strong mother-infant bonds and hand-rearing of modified infants would violate this bond, causing stress and derivational harm to surrogate mother and infants alike.

In the study it was reported that of the 179 embryos that were transferred to 59 surrogates to obtain 14 live monkeys, there were eight miscarriages and four full-term still births (from difficult labour presumably due to in vitro procedures). In the end the research team ended up with nine live monkeys with various successful disruptions of the dystrophin gene. From an animal welfare point of view this is significant 'collateral damage' (or contingent harm) and some of the procedures and unintended effects would likely be classified as 'severe'. In a follow-up study which examined both aborted tissue as well as two of the living monkeys (now 4 years old) using whole-genome sequencing no off-target effects were found (Wang et al., 2018), nothing was reported on the welfare of the surviving DMD monkeys.

4.5 CONCLUSION: ETHICAL AND EPISTEMOLOGICAL PERSPECTIVES

From the different outlooks on animal welfare, scientists and other personnel working with animals can try their best to safeguard animal welfare but the inherent uncertainty involved in genetic editing adds a hazardous dimension

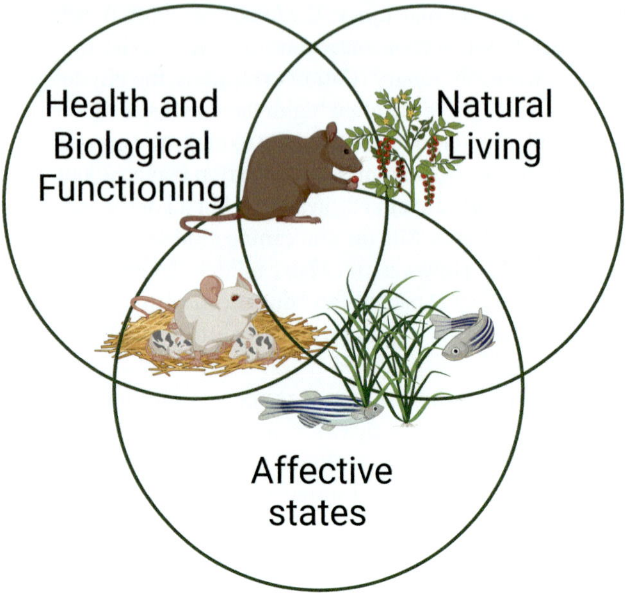

Figure 4.1 The three welfare paradigms. (Modified from Fraser et al., 1997. Created with BioRender.com.)

to this task. Do we have sufficient knowledge about what we are doing? Pleiotropic and off-target effects could result in animals with unintended welfare problems. In genetically altered animals, there might be increased difficulty assessing welfare and/or levels of discomfort/pain. Altered animals might have novel properties and display novel preferences and needs unknown to those who care for them. New animal models for human psychiatric and neurological disorders and diseases might inflict a new dimension of suffering as well as challenge our conception of the human/animal dichotomy as animals with humanized brain would be expected to further improve disease modelling.

Is the solution to create CRISPR animals better suited for laboratory life? Should welfare be ensured by engineering animals that are more trusting, docile, insensitive to pain and psychologically resilient or should good socialization, training and management be sufficient? Although it is scientifically within our grasp to do so, it will take generations of trials and many individuals for maintaining the genetic lines to achieve such animals and because they might have lost their natural behaviours and reactions, we might be unable to comprehend how their environment and the procedures we expose them to truly impact them (Figure 4.1, Table 4.1).

REFERENCES

Algers, B., Blokhuis, H., Boetner, A., Broom, D. M., Costa, P., Domingo, M. et al. 2009. Scientific opinion: General approach to fish welfare and to the concept of sentience in fish. *EFSA Journal* 954: 1–27.

Balcombe, J. 2016. *What a Fish Knows - The Inner Lives of Our Underwater Cousins*, 1st ed. London, UK: Oneworld Publications.

Baumans, V. 2005. Environmental enrichment for laboratory rodents and rabbits: Requirements of rodents, rabbits, and research, *ILAR Journal* 46, 2: 162–170.

Boissy, A., Manteuffel, G., Jensen, M.B., Moe, R.O., Spruijt, B., Keeling, L.J. et al. 2007. Assessment of positive emotions in animals to improve their welfare. *Physiology and Behavior* 92: 375–397.

Braitwaite, V. and L. O. Ebbesson. 2014. Pain and stress responses in farmed fish. *Revue scientifique et technique (International Office of Epizootics)* 33: 245–253.

Braithwaite, V., Huntingford, F. and R. Van den Bos. 2013. Variation in emotion and cognition among fishes. *Journal Agricultural Environmental Ethics* 26: 7–23.

Broom, D. 1991. Animal welfare: Concepts and measurement. *Journal of Animal Science* 69, 10: 4167–4175.

Butler, A. B. and W. Hodos. 1996/2005. *Comparative Vertebrate Neuroanatomy: Evolution and Adaptation*. New York: Wiley.

Chandroo, K. P., Yue, S. and R. D. Moccia. 2004. An evaluation of current perspectives on consciousness and pain in fishes. *Fish and Fisheries* 5, 4: 281–295.

Chen, Y., Zheng, Y., Kang, Y., Yang, W., Niu, Y., Guo, X. et al. 2015. Functional disruption of the dystrophin gene in rhesus monkey using CRISPR/Cas9. *Human Molecular Genetic* 24, 13: 3764–3774

Cooper, J. J. and G. J. Mason. 2000. Increasing costs of access to resources cause re-scheduling of behaviour in American mink (Mustela vison): Implications for the assessment of behavioural priorities. *Applied Animal Behaviour Science* 66, 1–2: 135–151.

Dawkins, M. S. 1983. Battery hens name their price: Consumer demand theory and the measurement of ethological 'needs.' *Animal Behaviour* 31: 1195–1205.

Dawkins, M. S. 1990. From an animal's point of view: Motivation, fitness, and animal welfare. *Behavioral and Brain Sciences* 13, 1: 1–9.

Dawkins, M. S. 2017. Animal welfare with and without consciousness. *Journal of Zoology* 301, 1: 1–10.

Devolder, K. and M. Eggel. 2019. No pain, no gain? In defence of genetically disenhancing (most) research animals. *Animals*, 9, 4: 154.

Duncan, I. J. H. 1993. Welfare is to do with what animals feel. *Journal of Agriculture and Environmental Ethics* 6: 8–14.

Duncan, I. J. H. and D. Fraser. 1997. Understanding animal welfare. In *Animal Welfare* ed. M. C. Appleby and B. O. Hughes, 19–31. Wallingford, UK: CAB International.

European Commission. 2020. *Report From The Commission To The European Parliament And The Council, 2019 report on the statistics on the use of animals for scientific purposes in the Member States of the European Union in 2015–2017*. Brussels. https://ec.europa.eu/info/sites/default/files/com-2020-16-f1-en-main-part-1.pdf (accessed August 29, 2022).

Farm Animal Welfare Council (FAWC). 1993. *Second Report on Priorities for Research and Development in Farm Animal Welfare*. Tolworth, UK: MAFF.

Foulkes, T. and J. N. Wood. 2008. Pain genes. *PLoS Genetics* 4, 7: e1000086. https://journals.plos.org/plosgenetics/article?id=10.1371/journal.pgen.1000086 (accessed August 29, 2022).

Fraser, D., D. M. Weary, E. A. Pajor, and B. N. Milligan. 1997. A scientific conception of animal welfare that reflects ethical concerns. *Animal Welfare* 6(3): 187–205. doi:10.1017/S0962728600019795.

Goodyer, J. 2022. Geneticists aim to resurrect the Tasmanian tiger within the next 10 years. *BBC Sciences Focus*, August 18, 2022. https://www.sciencefocus.com/news/geneticists-resurrect-tasmanian-tiger-within-ten-years/ (accessed August 29, 2022).

Grunewald, S. 2019. CRISPR's creatures: Protecting wildlife in the age of genomic editing. *UCLA Journal of Environmental Law and Policy* 37: 1.

Huntingford, F. A., Adams, C., Braithwaite, V. A., Kadri, S., Pottinger, T. G., Sandøe, P. et al. 2006. Current issues in fish welfare. *Journal of Fish Biology* 68, 2: 332–372.

Jensen, P. and F. M. Toates. 1993. Who needs 'behavioural needs'? Motivational aspects of the needs of animals. *Applied Animal Behaviour Science* 37: 161–181.

Jønholt, L., Bundgaard, C. J., Carlsen, M. and D. B. Sørensen. 2021. A case study on the behavioural effect of positive reinforcement training in a novel task participation test in göttingen mini pigs. *Animals* 11, 6: 1610.

Katnelson, A. 2018. What to watch for when analyzing mouse behavior. *Spectrum*. https://www.spectrumnews.org/news/watch-analyzing-mouse-behavior/ (accessed August 29, 2022).

Key, B. 2016. Why fish do not feel pain. *Animal Sentience* 3: 1.

Kirchhelle, C. 2021. *Bearing Witness, Ruth Harrison and British Farm Animal Welfare*. London, UK: Palgrave Macmillan.

Kornum, A. L., Röcklinsberg, H. and M. Gjerris. 2017, The concept of behavioural needs in contemporary fur science: Do we know what American mink (Mustela vison) really need? *Animal Welfare* 26, 2: 151–164.

Lawrence, A. B., Newberry, R. C. and M. Špinka. 2018. Positive welfare: What does it add to the debate over pig welfare? In *Advances in Pig Welfare; Herd and Flock Welfare*, ed. M. Špinka, 415–444. Duxford, UK: Woodhead Publishing.

Lee, C. J., Paull, G. C. and C. R. Tyler. 2022. Improving zebrafish laboratory welfare and scientific research through understanding their natural history. *Biological Reviews* 97, 3: 1038–1056.

Lim, K. R. Q., Nguyen, Q., Dzierlega, K., Huang, Y. and T. Yokota. 2020. CRISPR-generated animal models of Duchenne muscular dystrophy. *Genes* 11, 3: 342.

Mellor, D. J. 2017. Operational details of the five domains models and its key applications to the assessment and management of animal welfare. *Animals* 7, 8: 60.

Mellor, D. J. and C. S. W. Reid. 1994. Concepts of animal well-being and predicting the impact of procedures on experimental animals. In *Improving the Well-Being of Animals in the Research Environment*, eds. R.M. Baker, G. Jenkin, and D.J. Mellor, 3–18. Glen Osmond, SA, Australia: Australian and New Zealand Council for the Care of Animals in Research and Teaching (ANZCCART).

Panksepp, J. 1998. Affective Neuroscience: The Foundations of Human and Animal Emotions. Oxford: Oxford University Press.

Perles, Z., Moon, S., Ta-Shma, A., Yaacov, B., Francescatto, L., Edvardson, S. et al. 2015. A human laterality disorder caused by a homozygous deleterious mutation in MMP21. *Journal of Medical Genetics* 52, 12: 840–847.

Reardon, S. 2019. CRISPR gene-editing creates wave of exotic model organisms. *Nature* 568: 441–442. https://www.nature.com/articles/d41586-019-01300-9 (accessed August 28, 2022).

Remmel, A. 2021. CRISPR based gene therapy dampens pain in mice. *Nature* 591, 7850: 359.

Rollin, B. E. 2004. The ethical imperative to control pain and suffering in farm animals. In *The Well-Being of Farm Animals. Challenges and Solutions*, ed. G. J. Benson and B. E. Rollin, 3–19. Oxford, UK: Oxford Blackwell.

Rose, J. D., Arlinghaus, R., Cooke, S. J., Diggles, B. K., Sawynok, W., Stevens, E. D. et al. 2012. Can fish really feel pain? *Fish and Fisheries* 15, 1: 97–133.

Russell, W. M. S. and R. L. Burch. 1959. *The Principles of Humane Experimental Technique*. London: Methuen. Facsimile edition (1992). Potters Bar, Herts: UFAW.

Sauer, U. G. 2000. Reasons for not using primates in research. *ALTEX* 17, 4: 217–220.

Schultz-Bergin, M. 2018. Is CRISPR an ethical game changer? *Journal of Agricultural and Environmental Ethics* 31, 2: 219–38.

Sneddon, L., Braithwaite, V. and M. Gentle. 2003. Novel object test: Examining nociception and fear in the rainbow trout. *The Journal of Pain* 4, 8: 431–440.

Špinka, M. 2006. How important is natural behavior in animal farming systems? *Applied Animal Behaviour Science* 100: 117–128.

Stolba, A. and D. G. M. Wood-Gush. 1989. The behaviour of pigs in a semi-natural environment. *Animal Science* 48, 2: 419–425.

Tannenbaum, J. 1991. Ethics and animal welfare: The inextricable connection. *Journal of the American Veterinary Medical Association* 198, 8: 1360–1376.

Taylor, K. and L. R. Alvarez. 2019. An estimate of the number of animals used for scientific purposes worldwide in 2015. *Alternatives to Laboratory Animals* 47, 5–6: 196–213.

Thomasy, H. Cloning wildlife and editing their genes to protect them and us. *NEO.LIFE*, May 6, 2021. https://neo.life/2021/05/cloning-wildlife-and-editing-their-genes-to-protect-them-and-us/ (accessed August 18, 2022).

Wang, S., Ren, S., Bai, R., Xiao, P., Zhou, Q., Zhou, Y. et al. 2018. No off-target mutations in functional genome regions of a CRISPR/Cas9-generated monkey model of muscular dystrophy. *Journal of Biological Chemistry* 293: 11654–11658.

Yates, J. 2018. Naturalness and animal welfare. *Animals* 8, 4: 53.

Zuo, E., Cai, Y., Li, K., Wei, Y., Wang, B., Sun, Y. et al. 2017. One-step generation of complete gene knockout mice and monkeys by CRISPR/Cas9-mediated gene editing with multiple sgRNAs. *Cell Research* 27: 933–945.

5

Beyond welfare

MICKEY GJERRIS
University of Copenhagen

JAMES YEATES
CEO, World Federation for Animals

ANNA KORNUM
Independent Researcher

DORTE BRATBO SØRENSEN
University of Copenhagen

HELENA RÖCKLINSBERG
Swedish University of Agricultural Sciences

DOI: 10.1201/9780429428845-5

5.1 INTRODUCTION

Most people will agree that if I cause an animal to experience severe pain out of sheer boredom or out of a warped understanding of what constitutes 'entertainment', I am doing something wrong. It is this basic sense of something being 'wrong' that can be said to constitute the foundation of ethical experience. As humans we do not live in a universe where things are only factually right or wrong as in "the Earth revolves around the Sun", but also in a universe where things are ethically right or wrong as in "you should not steal from the poor". Not that I cannot steal. It is physically possible for me to do so, even though the consequences might not benefit me as I could get imprisoned or at least get ostracized by my peers. Society will pass an ethical verdict on my action. So, where actions that violate the laws of nature are impossible, actions that violate ethical rules are only all too possible.

As can be readily seen ethical rights and wrongs are very different from what we here will label 'scientific truths'. The latter is the subject of scientific inquiry following certain methodologies depending on the subject. We use different methods when we want to examine the feeding habits of hippos as when we want to establish the function of a certain protein in the saliva of hippos. Both are, however, nevertheless scientific inquiries where we, to the best of our ability and remembering the uncertainty that follows all scientific results, try to describe how the world *is*, while respecting Karl Popper's dictum that scientific hypothesis can never be proven, only falsified.

Ethical rights and wrongs are evaluations of what humans *do* based on an understanding of what they *ought* to do. Ethics in this sense is passing judgement and saying whether something is right or wrong. This also means that ethical reflection is only relevant for certain phenomena. The structure a certain protein has and the phagocytic mechanism of a leucocyte are some things we can be right or wrong about, but it seems odd to say that our belief regarding the matters be right or wrong in an ethical sense (exceptional cases such as intentional ignorance or biases aside). It would make no sense to say that protein molecules have this or that molecular composition or phagocytes have this or that biochemical-mechanical action, but we ought to think they have a different one.

It is different with humans, most of whom are *moral agents*, i.e., capable of – to some extent at least – understanding the consequences of their actions and capable of – to some extent at least – choosing what to do. Thus, we can not only scientifically describe humans' feeding habits and evaluate to what extent they fulfil their physical needs, but also evaluate them ethically and say whether they are right or wrong as seen from our understanding of how and what they ought to eat from an ethical perspective.

For example, discussions flourish around the so-called obesity epidemic in many parts of the world, as obesity can carry serious health issues for those affected. To some it is wrong to harm your own body and people should simply stop doing that. To others obesity is much more a result of external factors such as genetics or social and cultural factors. Others see the problem rooted in more structural economic reasons such as the globalized food industry (Mozaffarian, 2020). Some find that the ethical issue is that obesity has a negative effect on the individual's possibilities to live a flourishing life while others are concerned about the strain on public health costs (ten Have, 2014). And these are just some of the ethical issues surrounding human eating habits. As an example, the environmental consequences, especially related to climate change, are coming more and more into focus as is concerns for the welfare of the animals involved in food production (Gjerris, 2015, World Federation of Animals, 2023).

What can be learned from this is that for moral agents, ethics enters the picture when we find that something is right or wrong … and it could have been different, if different choices had been made. The lion is not a moral agent in the same way and does not do anything ethically wrong when she naturally hunts down the gazelle even though this obviously negatively affects the welfare of the gazelle. The lion does not have a choice. We can lament the suffering of the gazelle and ponder why pain and death is an integrated part of the universe. If we have religious convictions, we might even find it a tragedy. But we cannot blame the lion for what it is doing.

We can, however, debate whether it is ethically right for a human being to cause pain to an animal. Most, albeit not all, will agree that it depends on the situation. If my child is attacked by an animal and her life is threatened, few will disagree I am right in defending my child even if I cause pain to the animal. If my child out of boredom harms an animal, most will agree that I am right in stopping my child. But rarely are the situations we evaluate ethically this simple and rarely do we agree on what is right or wrong which can be readily seen if one just takes a cursory glance at the debate on farm animal welfare and ethics.

People working within the field of research animals are not unfamiliar with ethical discussions as these. Some are fundamental to the use of animals in research. Is it ever ethically acceptable to use animals for research purposes? If so, under which circumstances could it be ethically acceptable? How much suffering might it be ethically warranted to inflict on an animal? To what extent does the purpose or deliverables of the research play a role? Those are just some of the over-arching issues that frequently pop up in the debate and are highlighted throughout this book. In the following we will look at some of these issues through the perspective of animal ethics, understood as the critical and systematic reflection on how humans ought to treat animals.

5.2 ETHICS AS CRITICAL AND SYSTEMATIC REFLECTION

Most of us are prone to judge others ethically. When we see someone doing something that is wrong or experience human suffering that we find unwarranted, we judge. Every time you find that something is ethically right or wrong, you make yourself the judge. This can be an uncomfortable realization as most of us readily recognize that this entails that others do the same to us. And who are they to tell us what is right and wrong? The short answer is that they are the same as us. Beings that have the intelligence to discern whether something is right or wrong as seen from their perspective.

Our understanding of what is morally right and wrong in the specific situations that we find ourselves in is based on our ethical world view – a sense of what is right or wrong in general. This is not to place to go deeper into how our world views are constituted. It is enough to point to that we sometimes disagree. Sometimes on the more general notions of right and wrong and sometimes on how these should be interpreted in specific situations. Thus, some find that it is wrong to kill an animal and eat it, no matter how the animal has been faring, whereas others find this unproblematic as long as the animal has had a good life. The disagreement here is on the general level: is it ethically justified to kill an animal to eat it? Those that agree on this can, however, find themselves disagreeing on whether this particular production method offers the animals a good life. That is, they might disagree on the ethical evaluation of facts regarding housing and welfare parameters and on how to define 'good enough' when it comes to animal welfare.

In large as well as small cases, we find ourselves disagreeing on what is right and wrong. Ethics can help us to a better understanding of what we disagree about. Ethics can uncover the general values along which we try to organize our lives as well as shed light on the assumptions that go into our interpretation of the specific situations that we pass different judgements on.

But ethics is more than just trying to understand why we disagree, even though that can often be helpful enough and bring the discussion forward in a fruitful way. Ethics is also putting forth arguments supporting both general and specific judgements. This is not possible through the scientific methods used when trying to resolve a disagreement on how the world is, e.g., whether a certain vaccine has side effects or not. Rather, ethics uses various ways of argumentation structures to convince others that a certain world view is better (more right) than others and if agreed upon, they should, in the name of coherence, adopt this and thus pass the same judgements as the one making the argument. In Western philosophy, encompassing attempts to convince others about what should be considered right and wrong are often labelled 'normative ethical theories'.

5.3 NORMATIVE ETHICAL THEORIES

Through the winding road of human existence, a huge number of such theories have arisen. They share the ambition to exclaim something systematic about what is right and good in connection with human thought and action. Besides that, there is not much they agree on. How such an inquiry should take place, what role, e.g., experience and emotions should play, to what extent the theory should be understood as all-encompassing, to what extent universal ethical norms or principles can be formulated and, if so, are able to guide us, etc. are all very much disagreed upon. This is not the place to dive into all these disagreements or explain why it is so hard – some would say impossible – to come up with a definite ethical answer. We here refer you to existing textbooks on ethics written for non-experts, e.g., Gjerris et al. (2013).

Instead, we have asked five ethicists with different perspectives on ethics to present their standpoint and discuss three cases from the animal research world to show how differently ethical issues can be addressed. It should be noted that many more could have been included. We have chosen the ones most prevalent in the discussions on animal research in the Western world. As you will see the diversity is already thought-provoking and we urge you to look further than this book to encounter the full diversity of ethical thinking.

Here we will instead introduce some of the basic discussions within ethics of relevance to animal research, focusing on four issues: Who is of ethical importance? Is welfare all that matters? Does the death of animals matter? Do animals have integrity?

5.4 WHO IS OF ETHICAL IMPORTANCE?

It might seem obvious that the animals themselves are worthy of ethical consideration when discussing animal research ethics, but this has not always been the case. Within Western philosophy there has been a strong tendency to only view humans as ethically relevant. This way of thinking is often labelled 'anthropocentric' (from the Greek ἁνθρωπος, meaning 'human'). The roots for this can be found back in, e.g., the thinking of Aristotle (384–322), who argued that everything has a purpose, so that plants were created for the sake of animals and animals for the sake of humans, but that of all beings humans are the only ones who have a purpose in themselves. The main tendency within Christianity has also been to see humans as special and above all other created beings, often referring to the book of Genesis, chapter 1, where it is stated that humans (as opposed to everything else) are made in the image of God (imago dei) (White, 1967; Rauw, 2015). It should be noted, however, that the relationship between humans and animal has also been interpreted very differently within a Christian context, see, e.g., Preece (2005).

Later thinkers such as Descartes (1596–1650) argued that animals were just biological machines (automata) and their reactions to, e.g., being subjected to pain were mere reactions as when a clock strikes the hour. Immanuel Kant (1724–1804) went down a different road, but also ended up with animals being outside the realm of ethical consideration. To Kant what granted a being ethical importance in and of itself was its capacity for rationality, especially its ability to contemplate the difference between the ethically right and wrong. And as Kant only saw these capabilities in humans, they thus became the only beings that matters ethically speaking. It should be noted that Kant still opposes cruelty to animals, but on the grounds that it might lead to cruelty to humans. Animals are thus a sort of moral training ground (Steiner, 2005).

Today the idea that the pain and suffering of animals are not a legitimate part of ethical consideration has only few followers. Animal welfare science and human everyday experience point to animals being not machines but sentient beings. The idea that only rational beings should be included in the ethical community seems intuitively wrong to many. Not least since it by implication excludes a large number of humans from being considered ethically relevant in themselves, e.g., babies, severely demented persons, seriously mentally ill persons, persons in coma, etc.

One could argue that the ethical importance of humans is not based on specific capabilities, but rather on their biological kinship. Here the Australian philosopher Peter Singer (1946–) has famously argued that to include all humans in the ethical community, but not animals, only because that they are humans and animals are not, would be akin to racism or sexism. It would be excluding animals from the ethical community based on a non-relevant trait (species) without considering relevant similarities – that both humans and some species of animals are sentient. Consequently, he labels such positions as being an expression of speciesism.

Peter Singer infamously argued that if one is not willing to use a human with the same mental capacities as a dog in an experiment, then one ought not to use the dog. This initially provoked many people as they believed Singer was reducing the ethical importance of mentally disabled people, but the argument actually aimed for the opposite: if the amount of suffering is the same in a human and in a dog in any given experiment, there is no ethical ground for claiming that it is wrong to perform the experiment on the human, but not on the dog as the difference of species is ethically irrelevant. It is the amount of suffering that is important. Not to make the call based on that is to be 'speciecist', favouring one species over another based on an irrelevant parameter (species), instead of looking at what is relevant (suffering) (Singer 1975/2015).

For Singer and many other ethicists, the ability to experience one's own existence in terms of welfare (being sentient) is the relevant capacity that should be used to decide whom and what rational beings owe ethical consideration. The position is thus often labelled 'sentientism'. All humans and non-human animals that are sentient have the capacity to subjective mental experiences of their existence in terms of feeling pain, fear, contentment, joy, etc. and therefore belong to the ethical community.

This obviously complicates matters ethically as compared to the view that only other humans warrant ethical attention (and that can be complicated enough). First, one needs to develop an idea of which animals qualify as 'sentient'. For an in-depth discussion of this, please see the previous chapter. Here we will just assume, in line with current legislation in, e.g., North America and in the EU, that animals typically used in laboratories such as various mammals, birds and fish are sentient. An ongoing discussion within the animal welfare science community is whether some species of insects should also be considered sentient as some evidence points to that. We will not go further into this discussion here, but urge you to investigate the literature, if you are working with insects as matters are more complicated than most of us initially think, see, e.g., Gibbons et al. (2022).

That animals should be taken directly into the ethical considerations when using them in research is thus being seen as a sentientistic viewpoint. Whether one then shares Singer's viewpoint that the species in question is completely irrelevant and only the amount of welfare affected is important can be seen as a discussion within this overall viewpoint.

5.5 IS WELFARE ALL THAT MATTERS?

That welfare – understood broadly as subjective mental experiences of a positive or negative nature, health and biological functioning, and possibility to unfold species-specific behaviours – matters is widely accepted within the different ethical perspectives present in the literature and presented later in this book. Whether there are other concerns than welfare concerns that ought to play a role is, however, contested. The view that only welfare matters has been labelled 'welfarism' and is often combined with the idea that a certain amount of animal welfare loss is acceptable if it increases the overall welfare (Sumner, 1988). This way of thinking is probably very recognizable to people working within animal research where the negative animal welfare induced by the research is seen as justified by the potential welfare gains that the research can bring to either humans or – in some cases – other animals. For an elaboration of this view, please see Chapter 6.

As can be seen in Chapter 3 this is also the raison d'être between much animal welfare legislation, sometimes combined with what is known as 'side constraints' – limits as to how much welfare can be 'traded' around. Thus, the EU Directive on animal welfare in animal research, article 15 (2010/63/EU) explicitly states: "... *Member States shall ensure that a procedure is not performed if it involves severe pain, suffering or distress that is likely to be long-lasting and cannot be ameliorated*". There are thus seemingly non-welfare concerns that here play a role in the regulation of the area. However, article 55.3 clarifies that exceptions to this can be made if there are exceptional and scientifically justifiable reasons, although individual member states may decide not to allow for such exceptions, if the animals involved are non-human primates.

The view that welfare and only welfare is relevant for ethical considerations is, as mentioned above, sometimes labelled 'welfarism' and closely connected to the ethical perspective 'utilitarianism'. The view that other concerns ought to play a role as well can be found in many other ethical perspectives (Yeates et al., 2011). Traditionally the view has been attached to rights-based positions, often labelled 'abolitionist', as they oppose all instrumentalization of animals arguing that each individual ought to be respected as an end in themselves and not reduced as means to maximize the overall welfare. To dive deeper into this discussion and see how other positions than the rights-based ones (see Chapter 7) find the utilitarian 'welfarism' position problematic, please see Chapters 8–10 later in the book.

Regardless of ethical position, we can say that claiming that animal welfare matters implies the view that their welfare should be at least taken into direct consideration when research is being planned and performed. Different positions will then have different opinions on whether and to what degree the welfare of the individual can be traded for welfare gains for other individuals or whether this welfare trade is ethically acceptable at all (Röcklinsberg et al., 2017). Yet another question is whether there are other ethical concepts than welfare to consider.

An example from inter-human relationships can show the importance of this question: Is it OK to lie to another human being, if it increases the welfare of that person? We realize that this is a difficult question for many people to provide a simple answer to. Most of us would like to know the context, the amount of welfare at stake, etc. However, unless you find that telling the truth and telling lies are just equally neutral instruments to produce welfare, our guess is that you find that, all things being equal, one should tell the truth even though it might have negative welfare consequences for yourself or others. And then based on that initial reaction, consider whether the context could provide reasons to lie. Also consider that what usually happens is that we find it necessary to explain why we lied. Only rarely do we ourselves or others demand an explanation from us on why we spoke the truth. Hence, many of us have everyday experiences that point to that welfare is not all that matters when we consider ethical issues (Gjerris et al., 2013).

Here we will examine two possible phenomena that are relevant to animal research and that, as telling the truth, might be ethically relevant although they are not only relevant for the welfare of the animal: animal death and animal integrity. As with welfare the combination of animal research and biotechnology raises special issues and issues that have been present in animal research from the beginning. Here we will look first at the more general issues and then focus on the special challenges that the additional use of biotechnology raises.

5.6 DOES THE DEATH OF ANIMALS MATTER?

Death is an inevitability for all living things. Everything will come to an end. From the oak tree in my garden to the potted plant in my windowsill. Not breathing is as natural as breathing – and will be the state for much longer for each organism. “Life is postponed death” as the Danish ethicist K.E. Løgstrup said (Løgstrup, 2020/1956). As such it seems a bit weird that it is so important to us when it concerns our own death. Most of us would prefer for life to continue. But we are adult humans, and we know what we stand to lose. As far as we know that is not the case of (most) animals. They do not spend their time pondering the meaning of life in light of their approaching death, but just go unperturbed about their daily business. This can be seen as part of the explanation that many people tend to ascribe less importance to the death of animals than the death of ourselves or other humans. Thus, many people have concerns about the welfare of the animals they eat – how was their life, while they had it – but recognize, at least on some level, the fact that their lives were ended to enable them to eat those very same animals. To get a better understanding of how the death of animals could constitute an ethical issue, we have divided the discussion into three aspects: (A) The welfare of the process, (B) the potential loss of welfare due to being dead and (C) non-welfare issues with being dead. In doing so, we will here limit ourselves to discuss cases where animals die by the hand of humans so as not to complicate matters unnecessarily.

5.6.1 Death as a welfare issue: The experience of the animal

One way in which death seems important is that the experiences of the animal during the dying process are clearly a welfare issue. Insofar as it is impossible to die without a dying process, death is associated with the welfare issues involved in those processes. Two things are worth noticing here. Firstly, if an animal's dying process was somehow experientially pleasant (one might imagine some forms of opioid overdose), the process itself could include positive mental experiences. Secondly, if the animal on the other hand went through a very painful process but was then miraculously saved at the very point of death, we would still consider it a welfare issue. These concerns therefore appear to relate what is bad in the dying process in this perspective to the experiences during the dying phase. This suggests the dying process is not intrinsically bad, but merely contingently so.

Notwithstanding the above, we might add another concern. We sometimes seem to feel particularly concerned about the last experiences of an animal. We tend to consider the experiences of an animal before it dies as somehow more important than prior experiences, even if they are equivalent in terms of intensity, duration, etc. Positively, we might feel comforted that an animal rescued from cruelty 'died happy', even if its last moments were quantitatively dwarfed by a lifetime of suffering. Negatively, we might worry more about the moments before death, perhaps because they cannot be 'made up for'. This might explain why such a high amount of research has focused on 'humane slaughter/killing'. We might consider this as irrational when pondered upon or we might consider that this does reflect some deeper concern that should be respected. If the latter, then the dying process – because it is dying – is more important than experientially equivalent periods. Death has a retroactive effect on the value of the prior experiences.

In this context, we might be concerned with particular experiences at the point of or preceding death. One example is the worry of animals dying isolated from loved ones. While avoiding inappropriate anthropomorphism, it can still feel wrong that a (social) animal should die alone as when, e.g., a laboratory rat and her litter are killed separately, in addition to concerns for their stress during the separation.

Another way to view the value of death in relation to the dying process is to consider not only what death *involves* but what it *implies*. Even if, as a category or genus, 'dying' is not seen as necessarily or essentially bad, it contingently and accidentally often is. Put simply, an animal's death implies some dying process, which is likely to have been unpleasant. Broadening this idea out, we might also say that the quantity of death also signifies some prior badness, not only in the dying process but also the factors that contributed, or predisposed, to death. Some of these might well involve negative welfare states (e.g., research involving painful conditions/experiences). This gives us a reason to use measures such as mortality rate – or its reciprocal, longevity – as indirect welfare indicators of unpleasant experiences.

Taken together this means that the causes and processes of dying can be an ethical issue depending on the experienced welfare of the animal. In addition, our experience tells us that both the process leading up to the death of the animal and the death process itself should be in focus as welfare issues are known to be involved in both.

5.6.2 Death as a welfare issue disregarding the experiences of the animal during the process

Essentially, we might suggest that an animal's death – or, better, an animal's not being alive – represents the absence of certain experiences (Yeates, 2010). In brief, we might modally compare two lives – one longer and one shorter life and find that the difference in value between them relates to the value of the period of life that would occur in the former and not the latter. We might also think of death as precluding the opportunity for experiences (which also avoids challenges in determining counterfactuals).

The idea of humane endpoints as used in animal research helps make this clear. A rat who is expected to suffer, with no meaningful hope of enjoyable experiences, can be said to be 'better off dead' (or more accurately, better off stopping living) earlier rather than later compared to the life that she/he will otherwise be allowed. It is interesting that this everyday example is one where death is considered as better for the animal. So, it seems that death is generally accepted as beneficial relative to the realistic counterfactual. Therefore, it seems to make sense to claim that in a case where the welfare of the rat from above would be less impaired and the rat could be said to be 'better off living', death constitutes a welfare issue with regard to the future welfare that is denied the rat by killing it. Within this perspective life and death are neutral states. They can be instrumentally better or worse for the animal. Death, as the absence of life, is the negation of the value that that life had. So, if the quality of an animal's life was going to be overall good, death is worse for the animal. If the life was going to be overall bad, death is better. These ideas are often considered in terms of whether an animal's life is 'worth living' or 'not worth living' for the animal (Yeates, 2017).

Defining when an animal's life is worth living or worth avoiding is difficult, not only due to the difficulty of placing ourselves in the situation of the animal and experiencing how he/she is experiencing it, but also because of the challenges of aggregating evaluations of interacting experiences. Even when we seek to understand whether another human's life is worth living, there is controversy over how 'reliable' those evaluations are. In practice, there may be some considerable grey zones between lives we can confidently say are worth living or avoiding for the animal. Nevertheless, many of us feel those calls can at least sometimes be made (e.g., when prescribing euthanasia for a beloved family dog). Indeed, we might use the neutral 'value' of death as a heuristic comparator to evaluate the life: if the animal would be better off dead, then the life is worth avoiding. This is circular reasoning but might be practically and psychologically useful.

We might also broaden out this approach merely to say that life is a necessary precondition for positive experiences and consider death to be wrong whenever it leads to a net reduction in such experiences *regardless of who has those experiences*. The point here is that one animal's death might not always be considered wrong if it is linked to more life for another animal (where both would be equally enjoyable). In other words, on this logic, one animal can be 'replaced' if the overall utilitarian calculus is unchanged or improved – just as we might not care about losing one storage vessel if it is replaced by another (of equal or greater size).

A related response to concern about an animal's death might be that the animal *would not have existed at all* were it not for the researcher (indirectly, at least). One might even argue that the animal somehow 'owes' its life to the researcher, as part of a (completely imaginary) contract. While this argument seems to risk post hoc justification for almost anything, it is worth noting here that its logic is predicated on whether the animal benefits from its life. If the animal's life is worth avoiding, then there is no such 'contract', merely exploitation. Note that this does not mean that killing such an animal would always be wrong – indeed, it would be euthanasia if it prevents further suffering; it rather means that the whole enterprise was overall wrong.

When animals would have good, albeit shortened, lives worth living (good housing, enrichment, limited contingent or procedural harms, etc.), then it would seem better to be alive for that time than not at all. In such cases, the animal's death can, however, be expected to prevent future valuable life, so it may still feel morally problematic. We might also worry that while creating an animal on that contract would seem legitimate at the time of creation, where the appropriate prospective comparison seems to be comparing shorter and no life at the time of creation, when it comes to the appointed time to kill it, the relevant prospective comparison now seems to be between a longer and short life, suggesting that killing it is 'now' wrong since the animal could still be kept alive beneficially.

Taken together, this perspective on the death of the animal still focuses on the welfare of the animal, but now in a lifespan perspective. Life and death in themselves are seen as neutral states that are to be evaluated based on the possible consequences to the welfare experiences of the animal overall (Figure 5.1).

5.6.3 Death as an ethical issue disregarding the experiences of the animal during the process, albeit not a welfare issue

A third approach is to consider that death has ethical importance regardless of the victim's acute and possible future experiences. Here we can only sketch some of these ideas, recognizing each might be developed in more complex ways. We will focus on 'death in itself' that was discussed under Section 5.6.2 rather than 'the process of dying' discussed under Section 5.6.1.

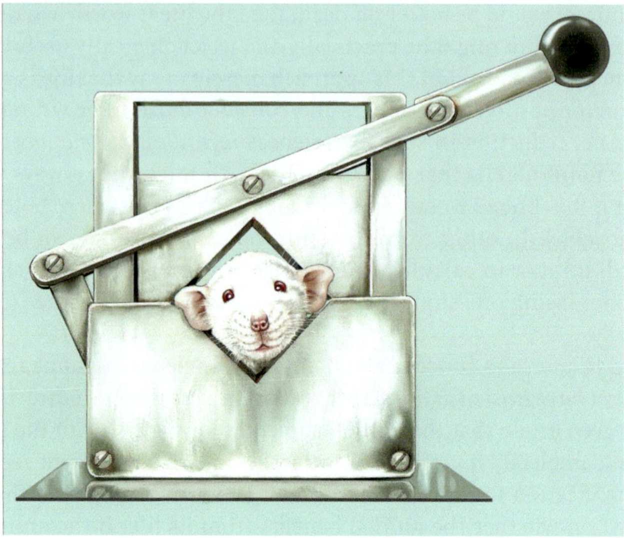

Figure 5.1 Guillotines designed for euthanizing small research animals such as mice and rats are standard in many places as it is considered a humane way of killing (or dispatching) the animals. Regardless of the potential welfare concerns, the issue of killing animals, of ending the life of an animal, raises ethical issues: Is life important in itself or only the experienced welfare while being alive? (Illustration Anna Kornum.)

Some will consider that killing an animal, no matter the reasons behind it, constitutes a wrong in itself. Killing is regarded as ethically problematic. The wrong is that the non-contingent value of life is disrespected. Most obvious here are concepts of a right to life or of sanctity of life. Such ideas consider that life has some value that should simply not be destroyed (if avoidable). One challenge for such ideas is how to manage their seemingly logical extension to *all* life. For example, if life has value independent of the experiences of the organism living this life, then this seems to apply to non-sentient lifeforms as well. Such approaches seem to endorse this as in the writings of Albert Schweizer (1875–1965) who coined the phrase "reverence for life" and found an appreciative attitude towards all living organisms to be the foundation of all ethical deliberation (Meyer & Bergel, 2002) or the environmental philosopher Paul Taylor's (1923–2015) idea of the *telos* of all living organisms as the basis for a *respect for life* and consequent rights for all life (Taylor, 1989).

Another road is to generate additional arguments as to why such reverence or respect is limited to only some beings, be it through an anthropocentric or sentientistic approach. The latter is often found in the animal rights literature as, e.g., in the work of Tom Regan (1938–2017), the most influential modern animal rights philosopher. Here the capacities of (some) animals and humans to be 'subjects-of-a-life' are seen as granting them a right to life as they are always to be treated as ends-in-themselves and never merely as a mean (Regan, 1983/2004).

This goes against the previously mentioned argument from Immanuel Kant (1724–1804), that we have only indirect duties to treat animals well because how we treat animals reflects on how we treat other human beings. However, as our experience tells us that most humans engage directly or indirectly in killing animals without necessarily mistreating other humans, at least in all the same ways, this argument seems to offer a weak reasoning for not killing animals. Surely – as rational beings – we can distinguish humans (as 'ends-in-themselves') from non-human animals (as mere 'things'). It should be acknowledged, though, that while theoretically this could be the case, as feeble partially rational beings, we could still be influenced in the way we treat humans in the same way that violent computer games are said to predispose players to non-virtual wrongdoing or pornography can lead to sexual objectification or sexism. It should be noted however that the empirical evidence also in these examples is inconclusive (American Psychological Association, 2020, Andrie et al., 2021).

5.6.4 Closing remarks

Finally, we might think of the symbolism of animal death. Morally differentiating human and non-human death in any way might seem to be evidence for a more widespread speciecistic attitude. Morally disregarding non-human animal death might seem to evince their exploitation. Indeed, the structures around which our morals are constructed – our laws, cultures, dogma and economies – are ones that include both a disregard for animal death and a regard for human benefits based on animal suffering and death. Examples of this are the continued use of animals in medical research in areas where well-founded criticism about the supposed benefits have been raised (Akhtar, 2015) and the widespread acceptance of 'surplus animals' (animals that are produced for the animal research sector, not least when using or breeding genetically modified animals, but for various reasons never used and subsequently killed) as a necessary part of animal research (Wewetzer et al., 2023).

Animal death can be seen purely as a welfare issue, both in the short and long terms, or as a wrong in itself. When we turn our attention to the specific issue of using biotechnology to produce animals for research, it is hard to point to any issues that are special for this way of utilizing animals for human benefits as compared to other ways. Animals die whether they are bred for research or agriculture, and they die whether they have been bred through traditional methods or through, e.g., CRISPR-Cas9.

One can speculate whether the breeding process causes fewer or more deaths when breeding the specific animal with the desired traits (e.g., a predisposition to a type of cancer) through traditional methods or biotechnology. Further it can be speculated whether the new breeding technologies in the end will give rise to fewer or more animals used overall as the precision of the new technologies might lower the number of animals in the individual experiments, or as 'surplus' but increase the number of experiments because of new opportunities to create, e.g., disease models that were not possible before.

However, it is hard to see that the issue of death itself changes when we move from one technology to another. As described above ethical perspectives will differ in what (if any) aspects of animal death that are seen as ethical issues.

5.7 DO ANIMALS HAVE INTEGRITY?

Integrity is one of these weird words that most of us connect with something positive but can have a hard time explaining what it actually means. The word comes from the Latin 'integer', meaning 'whole or complete'. When used about humans we often refer to the moral qualities in a person, the sense of incorruptness in an individual. A person with integrity is a person that stands by their values even at costs to themselves (e.g., telling the truth even though lying would have been more beneficial to them). When used about buildings we refer to the structural soundness of the house. A building with integrity will not collapse in the next autumn storm. But what do we mean when we speak of the 'integrity of an animal'? Or even the integrity of an entire species?

Initially it might be understood as the animal's ability to be the kind of animal that it is. Can it maintain its health and species-specific behaviour or are there outside factors that prevent this? Or, have the animals lost certain capacities due to being bred for specific purposes (production, research, etc.) (Rutgers & Heeger, 1999). Since this is a book on the ethical issues arising when using recent biotechnologies on research animals, we will focus our discussion on the latter question to examine whether these technologies can be said to violate the genetic integrity of the animals. In this sense the idea also ties closely to the original meaning of the word: 'whole or complete' and the question can be rephrased to: Do we violate the integrity of animals when we modify their genetic makeup through biotechnology?

5.7.1 The difference between naturally occurring and human-induced changes

If you in general adhere to the scientific theory of evolution (which we see no reason not to do) and have just a basic grasp of molecular biology, you already know that the animals discussed in this book (mammals, fish, birds, reptiles, insects, etc.) are all a result of genetic changes. Each species is the result of millions of years of evolution. So genetic changes happen naturally and cannot per se be said to be ethically problematic. At this level, the changes that occur in species over time are a scientific fact, just like the feeding habits of hippopotamus mentioned in the beginning of this chapter, but not a subject for ethical scrutiny.

Where changes in the genetic makeup of animals become ethically relevant is thus when they are induced by humans. This can happen on purpose as in selected breeding for desired traits in production animals (e.g., higher milk yield from cows) or as an unintended side-effect of human activities as the peppered moths in England during the 19th century that changed colour from light to dark due to increasing air pollution caused by the Industrial Revolution (Rice, 2007).

As the genetic changes in research animals are either a result of direct intervention to produce exactly this change or an unintended side-effect of an attempt to create another trait, we will here treat these as similar cases and limit our discussion to genetic changes that are directly induced by humans with the purpose of producing an animal that can be used in research, whether the actual result is then suitable for the intended research or can be used in other research.

What if, anything, could be wrong with this? Here the term integrity becomes important as it signifies that what was before and what is now are not the same. Something has, from a scientific perspective, changed on the molecular and perhaps phenotypical level and from an ethical perspective this change is what can be said to violate the integrity of the animal.

However, if one wishes to claim this as ethically wrong, something that should not be done, several counterarguments can be brought forth. We will introduce some of them here but be warned that each of them warrants a longer discussion than we are able to present here. The following should, however, be enough to get an idea of the complexity of the issue. For a fuller discussion of the concept of integrity in relation to biotechnology, we recommend Röcklinsberg et al. (2014), Hauskeller (2016) and Eriksson et al. (2018).

5.7.2 Integrity as a 'more-or-less' concept

Humans have bred animals (and plants) with specific purposes for approximately 15,000 years (McHugo et al., 2019). If it is claimed that using biotechnological tools to change the genetic makeup of research animals violates the integrity of the animals, it entails either the view that all other changes have been violations of integrity as well or warrant further arguments that show why only some changes should be seen as violating integrity. The latter can be based on several distinctions as, for example, the means used to induce the changes from simply looking at a group of animals and deciding based on this who should have offspring together, to more controlled conventional breeding, to using genomics to pair the animals at a molecular level to actually modifying the genome of the animal through, e.g., CRISPR-Cas9.

Here the tool that induces the change is in focus and one could claim that the degree of biological control decides the degree to which the integrity of the animal is violated. One could say that the less chance and the more human purpose the animal is an expression of, the less is left of the integrity of the animal. This obviously entails that integrity is not an 'either-or' but can be seen as a 'more-or-less'. In this sense the recent biotechnological possibilities can be seen as a step further down a wrong road, where the integrity of the animals is violated even more than previously. And here the wrong road can then be understood as a road where the animals to a growing degree are reduced from being whole and independent of humans (having integrity) to instruments for purposes that they have no interest in.

Finally, it would be possible to argue that two wrongs will not make it right, meaning that even though human history is littered with breeding of animals this does not justify continuing this with new and even more powerful tools, but should rather make us reflect on how to get rid of the practices already in use and seek to right the wrongs already made.

5.7.3 But where is integrity situated?

Changing the genome induces changes in a specific animal typically at a very early stage in the foetal process. Thus, in the case of this particular animal the end result is different from what would have been the result had there not been a human intervention. However, when this animal then has offspring, the question thus becomes whether these second-generation animals (and subsequent offspring) can be said to have had their integrity violated based on the argument that they would not have their specific genetic set-up without human intervention?

In the case of the originally modified animal there is an actual genome that was then changed. But regarding the offspring there is no actual animal that has had its genome changed. Seen this way the violation of integrity becomes a one-time event rather than something that is carried with in subsequent generations. This discussion might seem a bit airy, but it is a good example of what happens when a complex philosophical concept such as integrity is used in biotechnological discussions. The animals, the technology, the induced changes are all physical entities. They are present in a scientific sense, whereas integrity is an experiential phenomenon that we cannot point to in a physical sense or location, but that some nonetheless claim is present and can be violated. Pinpointing exactly where this violation takes place can, except for the case of the 'founding animal', be difficult.

The animal ethicist Bernard Rollin (1943–2021) argued that we should respect the telos of individual animals understood as their ability to live species-specific lives where they can unfold their behaviour and instincts. Here we interpret this to mean that not respecting the telos of the animal is the equivalent of violating its integrity. Rollin argued that only individual, actual, animals could have their telos respected (or not respected). Thus, he argued, it is ethically justifiable to change the telos of a species or sub-species of animals through genetic modification to another telos if the new telos is respected with regard to the novel individual animals (Rollin, 2015). In this way, Rollin avoids some of the unclarity of what is violated when. There are only individual actual animals and their telos to cater for. How they came about to have that telos becomes irrelevant.

Thus, it is a difficult task to pinpoint what exactly the ethical 'wrong' is in altering the genetic makeup of a species. First, the concept 'species' is a difficult concept, and it is very much discussed whether species actually exist, or if they should rather be understood as a heuristic tool developed by humans to better organize the world (Dupré, 1999). Further and as already mentioned, genetic changes happen quite naturally through the evolutionary mechanism of random mutation. All species constantly change to adapt to a changing environment. One can then make the argument more specific and claiming that integrity is only violated, when it is caused by direct (e.g., biotechnology) or indirect (e.g., climate change) human action, leaves open the question why human actions should be considered unnatural as compared to naturally occurring mutation or changes in the environment.

A possibility here could be to point to the changes induced as an expression of disrespect for the species in question, whether it is changed on purpose or as a side-effect to some other human goal-based action. A specific species could then be seen as a snapshot of a long and continued development based on random mutations, whose integrity is violated, if the changes are a result of actions performed by moral agents whose goals are external to the species in question.

This concern also shows how the concept of integrity is linked to the concept of 'naturalness' in a normative sense That is, what is natural is 'good' or 'right' and the opposite is true for something that is regarded to be 'unnatural'. Naturalness is, however, a notoriously difficult issue to apply to biotechnology and has been the concept of much debate. Some find that the distinction between 'natural' and 'unnatural' is meaningless as it implies that what humans do in some cases can be seen as unnatural even though humans themselves have grown put of nature. Others argue that certain technologies and/or approaches to animals and the end-results of these can rightly be described as 'unnatural' and often for that reason be evaluated as ethically problematic. Some of the discussions regarding the concept of integrity in relation to GM animals can be seen as almost exact copies of discussions about the relevance of the concept of naturalness. For a more thorough discussion of the concept of naturalness in relation to GM animals, please see Gjerris (2012).

Finally, integrity can be said to exist at the experience level of the individual animal. Here 'integrity' can be understood as an ideal most visible in the encounter with a wild animal. Here most will readily admit that there is an experience of wholeness, completeness, independence in the animal. It is so to speak 'done' before we arrive at scene. He or she has no use for us, and we have not (yet) turned it into something we can use. It simply is. Or as Röcklinsberg et al. (2014) write:

> 'Integrity' can thus be interpreted as the experience that an animal is whole, complete, and finished when we encounter it. Consequently, we cannot add integrity, but only take away from it. The integrity of the animal is thus violated when the animal is designed to serve human needs instead of being left to develop or fulfil its own species-specific goals. We have reduced it to an expression of our intentions and needs – we have de-animalized the animal. Something has been denied this animal. It is not aware of this, but it has nonetheless had its integrity violated, because it could have been allowed to come into existence, free to follow its own agenda, not ours.

5.7.4 Closing remarks

All these different aspects of integrity can, as can hopefully be seen, be discussed. They will carry more or less (if any) weight within different ethical perspectives and different understandings of the concept of integrity. What we would like to stress here is that, although perhaps tempting from a scientific perspective, the issues related to the concept of genetic integrity in research animals cannot be dismissed through science-based arguments. Obviously uninformed understanding of the nature of the science behind the technologies and the technologies themselves should be improved, but this will not remove all concerns about violations of animal integrity. Discussions needs to be held in a way that respects that we have other ways of understanding the world than just science and that these perspectives might also have something relevant to bring to the debate.

5.8 SO… WHERE DOES THIS LEAVE US?

We do not wish here to conclude whether death and integrity are ethical issues or not or conclude to what extent they should play a decisive role in ethical deliberations on the use of biotechnology in animal research. What we hope to have done is to raise awareness that there can be good reasons to have an open mind. And recognize that other issues than welfare can play a role and that there can be arguments both for and against including them. So, dismissing them out of hand and reducing arguments along these lines as irrational or unscientific is not a valid approach. We realize that it is often easier for people with a scientific background to relate to measurable and quantifiable welfare issues, but what we hope to have shown is that even these issues contain ethical concepts and evaluations that are not based on science and that other issues, for example, the issues of death and integrity are also relevant to contemplate and discuss when entering into ethical discussions on the use of biotechnology in animal research.

REFERENCES

Akhtar, A. 2015. The flaws and human harms of animal experimentation. *Cambridge Quarterly of Healthcare Ethics* 24, 4: 407–419.

American Psychological Association 2020. *APA Resolution on Violent Video Games*. Washington, DC: American Psychological Association.

Andrie, E. K., Sakou, I. I., Tzavela, E. C., Richardson, C., and A. K. Tsitsika. 2021: Adolescents' online pornography exposure and its relationship to sociodemographic and psychopathological correlates: A cross-sectional study in six European countries. *Children* 8, 10: 925.

Dupré, J. 1999. Are whales fish? In *Folkbiology*, ed. J Medi, and S. Atran, 461–476. Cambridge, MA: MIT Press.

Eriksson, S., Jonas, E., Rydhmer, L., and H. Röcklinsberg. 2018. Invited review: Breeding and ethical perspectives on genetically modified and genome edited cattle. *Journal of Dairy Science* 101, 1: 1–17.

Gibbons, M., Crump, A., Barrett, M., Sarlak, S., Birch, J., and L. Chittka. 2022. Can insects feel pain? A review of the neural and behavioural evidence. *Advances in Insect Physiology* 63: 155–229.

Gjerris, M. 2012. Animal biotechnology: The ethical landscape. In *Designer Animals. Mapping the Issues in Animal Biotechnology*, ed. C. Brunk, and S. Hartley, 47–70. Toronto: University of Toronto Press.

Gjerris, M. 2015. Willed blindness: A discussion of our moral shortcomings in relation to animals. *Journal of Agricultural and Environmental Ethics* 28, 3: 517–532.

Gjerris, M., Nielsen, M. E. J., and P. Sandøe. 2013. *The Good, the Right and the Fair. An Introduction to Ethics.* London: College Publications.

Hauskeller, M. 2016. *Biotechnology and the Integrity of Life.* London: Taylor & Francis.

Løgstrup, K. E. 2020/1956. *The Ethical Demand. Translated With an Introduction and Notes by Bjørn Rabjerg and Robert Stern.* Oxford: Oxford University Press.

McHugo, G. P., Dover, M. J., and D. E. MacHugh. 2019. Unlocking the origins and biology of domestic animals using ancient DNA and paleogenomics. *BMC Biology* 17: 98.

Meyer, M., and K. Bergel. 2002. *Reverence for Life. The Ethics of Albert Schweitzer for the Twenty-First Century.* Syracuse: Syracuse University Press.

Mozaffarian, D. 2020. Dietary and policy priorities to reduce the global crises of obesity and diabetes. *Nature Food* 1: 38–50.

Preece, R. 2005. *Brute Souls, Happy Beasts and Evolution: The Historical Status of Animals.* Vancouver: University of British Columbia Press.

Rauw, W. M. 2015. Philosophy and ethics of animal use and consumption. *CAB Reviews* 10, 16: 1–25.

Regan, T. 1983/2004. *The Case for Animal Rights.* Oakland: University of California Press.

Rice, S. A. 2007. *Encyclopedia of Evolution.* New York: Facts on File.

Röcklinsberg, H., Gamborg, C., and M. Gjerris. 2014. A case for integrity: Gains from including more than animal welfare in animal ethics committee deliberations. *Laboratory Animals* 48: 61–71.

Röcklinsberg, H., Gjerris, M., and I. A. S. Olsson. 2017. *Animal Ethics in Animal Research.* Cambridge: Cambridge University Press.

Rollin, B. E. 2015. Telos, conservation of welfare, and ethical issues in genetic engineering of animals. *Current Topics in Behavioral Neuroscience* 19: 99–116.

Rutgers, B., and R. Heeger. 1999. Inherent worth and respect for animal integrity. In *Recognizing the Intrinsic Value of Nature*, ed. M. Dol, M. F. van Vlissingen, S. Kasanmoentalib, T. Visser, and H. Zwart, 41–53. Assen: Van Gorcum.

Singer, P. 1975/2015. *Animal Liberation. A New Ethics for Our Treatment of Animals.* New York: HarperCollins.

Steiner, G. 2005. *Anthropocentrism and Its Discontents: The Moral Status of Animals in the History of Western Philosophy.* Pittsburgh, PA: University of Pittsburgh Press.

Sumner, L. W. 1988. Animal welfare and animal rights. *Journal of Medicine and Philosophy* 13, 2: 159–75.

Taylor, P. W. 1989. *Respect for Nature. A Theory of Environmental Ethics.* Princeton: Princeton University Press.

ten Have, M. 2014. Ethical aspects of obesity prevention. *Best Practice & Research Clinical Gastroenterology* 28, 2: 303–314.

Wewetzer, H., Wagenknecht, T., Bert, B., and G. Schönfelder. 2023. The fate of surplus laboratory animals. *EMBO Reports* 24: e56551.

White, L. Jr. 1967. The historical roots of our ecological crisis. *Science* 155: 1203–1207.

World Federation of Animals. 2023. *Unveiling the Nexus. The Interdependence of Animal Welfare, Environment & Sustainable Development*. Boston, MA: World Federation of Animals.

Yeates, J. W. 2010. Death is a welfare issue. *Journal of Agricultural and Environmental Ethics* 23: 229–241.

Yeates, J. W. 2017. How good? Ethical criteria for a 'good life' for farm animals. *Journal of Agricultural and Environmental Ethics* 30: 23–35.

Yeates, J. W., Röcklinsberg, H., and M. Gjerris. 2011. Is welfare all that matters? A discussion of what should be included in policymaking regarding animals. *Animal Welfare* 20: 423–432.

6

A utilitarian perspective on animal experimentation involving genetic modification

TATJANA VIŠAK
University of Mannheim

6.1 INTRODUCTION

In this chapter, I will introduce a utilitarian perspective on animal experimentation involving genetic modification. For this purpose, I will first make some general remarks about ethics and moral theories (Section 6.2). Then I will introduce the basics of utilitarianism (Section 6.3). Lastly, I will point out the considerations that guide a utilitarian evaluation of specific cases (Section 6.4). A utilitarian perspective might seem easy and straightforward, since it simply asks us to do whatever has the best consequences. Yet, its application to real-life cases, as we will see, is extremely complex, due to the complexity of real life.

6.2 ETHICS AND MORAL THEORIES

As humans we have extraordinary and rapidly increasing possibilities to manipulate and change the world around us. The influence of our actions may be far-reaching both in space and time. We can now more than ever deliberately affect

DOI: 10.1201/9780429428845-6

our genetic material and thus influence in novel ways who we are and who we may become. We can thus change the very nature of ourselves and of other individuals. As humans, we are also, as far as we know, the only beings that can deliberate about whether it is a good idea to do these things. We can explore our normative reasons for action and adjust our behaviour accordingly.

Ethics deals, among other issues, with normative reasons for action. We can talk about 'reasons' in different ways. When I tell some child that he had no reason to beat his brother, the child may reply that he had a reason. The reason for beating his brother was that he wanted to hurt him, because he was angry with him. If the child gives me this response, he states his *motivating* reason for action. But when I said that he had no reason to beat his brother, I meant that the action was not morally justified. Normative reasons for action are the facts that count, for the agent in question, in favour of performing the action. This does not necessarily imply that the agent is *aware* of the facts that count, for him, in favour of performing the action. Even if he is not aware of the normative reasons for or against an action in a particular situation, the reasons are still there. For example, if someone whose life I could easily save is in danger, I may have a normative reason to save this individual, even if I am not aware of it. The simple fact that performing a particular action *would save a life* may count, for me, in favour of performing the action. If I can perform either action A or action B, it may be the case that some facts count, for me, in favour of performing each of the actions and some other facts may count, for me, against performing each of the actions. For example, each action may have both good and bad consequences. In such a case, I need to know what I have *overall* reason to do. When I speak about 'reasons for action' without further qualification in this chapter, I refer to what we have all-things-considered normative reason to do. Ethics, as I see it, concerns, among other things, what we have reason to do. In that sense, ethicists explore whether actions are normatively 'justified', 'allowed' or 'forbidden'.

Ethicists want to know what we have reason to do and *why* we have reason to do (or not to do) particular things. For that purpose, ethicists construct, apply and evaluate *normative theories*. These theories aim at telling us in a coherent and systematic way what we have overall normative reason to do and why. Different normative theories tell different stories about what provides or grounds normative reasons for action. For example, according to *consequentialist* normative theories, only the value of the outcome that an action brings (or would bring) about is relevant for determining whether we have reason to perform the action. In contrast, according to *non-consequentialist* theories, whether we have reason to perform an action does not – or not only – depend on its consequences. These latter theories may hold that whether an action is 'right' or 'wrong' depends also or only on what kind of action it is or on the intention with which it was (or would be) performed. In this chapter, I will evaluate actions and practices from a consequentialist perspective. Thus, I will assume that the value that my action brings about is the sole determinant for what I have normative reason to do.

6.3 UTILITARIANISM

For consequentialists, an action's consequences matter, but what exactly are an action's *consequences* and which of them are relevant? In this chapter, I want to focus on *utilitarian* normative theories, which are by far the most prominent consequentialist theories. In fact, utilitarianism counts as one of the two classical theories in ethics (besides non-consequentialist Kantian theories, also known as deontological or rights theories) and it is still among the most popular ones. The ideas behind utilitarianism are simple and straightforward. Whether we have reason to perform an action depends on what the action brings about. Specifically, how does the overall course of the world, when the action is performed, compare to the overall course of the world, when an alternative action (including, perhaps, no action at all) is performed. We should perform the action that brings about the *best* state of the world. The best (or 'optimific') action is the one that maximizes value. I do not always explicitly mention actions and omissions, but it should be clear that it does not matter, according to utilitarianism, whether I bring about some consequence by doing something or by doing nothing. Since utilitarianism is 'maximizing' and since it does not recognize any morally relevant difference between doing and allowing to happen, it is very demanding (Sobel, 2007). There is no way of keeping one's hands clean by doing 'nothing' if one could have done something better instead.

The value that matters for utilitarians is welfare. Thus, utilitarians are *welfarists* about reasons for action (Crisp, 2006). They hold that welfare and only welfare provides reasons for action. 'Welfare' refers to well-being: the value that indicates an individual's quality of life. It is controversial what exactly makes an individual's life good or bad for her. Different theories of welfare propose rival answers to that question. For example, hedonists hold that lives are better to the extent to which they contain more pleasure (or positive mental states) and less suffering (or negative mental states). Thus, according to hedonists about welfare, all other things are good or bad for individuals only to the extent to which they promote pleasure or bring about suffering. These other things are only derivatively (or instrumentally) good or bad. The only ultimate good is pleasure and the only ultimate bad is suffering (Shafer-Landau, 2012, pp. 21–40). Preferentialism is an alternative theory of welfare, which holds that an individual's life is better for her, to the extent to which more of her preferences are fulfilled and fewer or her preferences are frustrated. To say it in simple words: an individual's life is better to the extent to which she gets what she wants (Shafer-Landau, 2012, pp. 42–56). Yet another theory of welfare holds that individuals are better off to the extent to which they live according to their nature and fulfil their natural potential. This kind of nature-fulfilment theory is also called 'perfectionism' (Sandoe et al., 2016, p. 62). This brief sketch of possible accounts of welfare should suffice for our purposes. No matter what exactly welfare consists in, utilitarians hold that the right action is the one that maximizes it.

If we are to maximize welfare, we need to consider the consequences of our actions in terms of welfare and we need to take into consideration the welfare of all affected individuals on an equal basis. My own welfare, or the welfare of my nearest and dearest ones, according to utilitarianism, does not count for more than that of others. This shows that utilitarianism as a moral theory is strictly impartial. What matters is the amount of positive and negative welfare, or of benefit and harm, that one brings about, but it does not matter *who* benefits and *who* is harmed. In that respect, individual identities are irrelevant when it comes to what we have normative reason to do. For example, it doesn't matter whether a cow gains five units of welfare and I lose one or whether I gain five units of welfare and the cow loses one: the overall effects on welfare would be the same and this is all that matters, according to utilitarianism.

But how can we 'count' welfare in the first place? Some may doubt whether welfare is the kind of thing that can be counted. But in fact, we do assess welfare in our daily lives. We do so when we ask ourselves how much we would benefit from different ways of spending our money or our vacations. We also assess the welfare of those near and dear to us and may distribute our attention accordingly. We compare welfare across time, when we say that our friend is better off these days than he was a while ago. We compare it across people, when we judge that one of our kids is better off these days than the other, who is having a difficult time at school. We even compare welfare across species, when we assess that a hen in a laying cage is worse off than a cow on a meadow. Of course, we may get it wrong in some cases, but in principle, welfare is the kind of thing that can be assessed and compared.

Some calculations are easy and intuitive, others are more complicated. For example, some actions may affect many individuals in completely different ways. The production of cheaper meat may have clear negative consequences in terms of more animal suffering. But it may be hard to assess the more indirect good or bad consequences that the production of cheaper meat may have. In daily life, some rough estimation will usually do, for example, when we decide how to distribute our attention among our kids or where to spend our holidays. But for some purposes more specific calculations are in order. For example, in a public health context it is common to calculate the welfare consequences of alternative projects to spend scarce resources most effectively. For these purposes, specific constructs, such as 'quality-adjusted-life-years' are in use (Torrance and Feeny, 1989). Sometimes it may be impossible to know which of our actions brings about the best consequences. This challenge is referred to as *the cluelessness problem* within utilitarianism (Lenman, 2000). In these cases, utilitarianism may still correctly tell us that whatever brings about the best consequences is the right thing to do, but that information does not help us to guide our actions, due to our limited knowledge about our actions' consequences.

As I already mentioned about utilitarianism's impartiality: for utilitarians a particular amount of suffering or enjoyment counts the same, no matter who experiences it. In that sense utilitarians reject any form of discrimination, for example, based on gender, race, or species. A particular amount of pain experienced by a mouse counts the same as the same amount of pain experienced by a human. Peter Singer, the most influential utilitarian animal ethicist, calls this feature of his theory 'anti-speciesism' (Singer, 2011). It allows for an impartial consideration of welfare. (I speak of pain and suffering here, because classical utilitarians were hedonists and because these are important considerations even according to non-hedonistic theories of welfare). In any case, what matters is the *overall amount* of good and bad that an action brings about. It does not matter who experiences it and how it is distributed.

There is some controversy among utilitarians about what exactly the requirement to maximize welfare entails. Are utilitarians ultimately interested in improving lives? Or are they interested in the overall amount of welfare in the universe? In many cases there is no need to decide between these views. If I make someone better off, I improve his life and, all else being equal, I also thereby increase the amount of welfare in the universe. But there are ways of increasing the amount of welfare in the universe that do not make anybody better off. If I bring additional well-off individuals into existence, I do not make anybody better off. After all, if the individuals never existed, they wouldn't be worse off. They would not 'be' in any way at all. Yet, by bringing well-off individuals into existence, I do, all else being equal, increase the amount of welfare in the universe (Višak, 2013).

These different utilitarian views about what ultimately matters have different practical implications. If I hold that I do something good by bringing additional well-off individuals into the world, this good may outweigh some bad. Some utilitarians accept the *replaceability argument*. They hold that, e.g., killing an animal that would otherwise have had a good future can be allowed, if I replace this animal by a 'new' animal that would not otherwise have existed and whose life is at least as good as the future of the killed animal would have been. In contrast, those who care about improving lives would not agree that killing and replacing animals in that way is okay. After all, one makes a life worse by depriving an animal of its good future and one doesn't improve any life by bringing an additional animal into existence.

How good a life is, according to utilitarianism, depends on the amount of welfare that it entails. According to the standard account of harm and benefit, the counterfactual, comparative account, some event benefits me just in case it makes me better off overall than I would otherwise have been. So, one compares my lifetime welfare in the world in which the event takes place with my lifetime welfare in the counterfactual world in which the event had not taken place. Here again, one might need to base one's judgement on informed guesses, since one may not know for sure what would have happened had the event not taken place. According to that account of harm and benefit, dying earlier rather than later harms me just in case my life would otherwise contain more welfare. This is the so-called *deprivation view* about the harm of death (Bradley, 2009). It explains why premature death is normally bad for individuals and why dying earlier rather than later can be the better option if the future of the individual would have been bad rather than good for it. If two individuals have different lifespans, their welfare at each point in time may be the same. Still, the individual that lives longer has more lifetime welfare, at least if one accepts that standard view, according to which lifetime welfare consists in the sum of temporal welfare.

With these basics about utilitarianism in place, I will now show how utilitarians go about evaluating specific cases, such as the cases involving animal experimentation and genetic modification that were presented in this book.

6.4 EVALUATING THE CASES

Utilitarians cannot say and do not say that animal experimentation or genetic modification of animals is always right or always wrong. Whether or not it is justified to allow a particular experiment or, for that matter, the practice of animal experimentation depends on the consequences of the experiment or the practice in question. These consequences, which need to be assessed in terms of the welfare of all concerned individuals, need to be compared to the consequences of alternative actions that are available to the agent in question. In principle, the balance of consequences might turn out either way, for or against the action. How it turns out depends on many details, such as:

a. What are the effects of the action in terms of welfare? Are they positive or negative and to what extent? (Here, we need to assume some account of what welfare consists in, i.e., some theory of welfare.) In practice, we will often have to work with very rough estimations of an action's effects on welfare, since we do not have all the necessary data available.
b. How do these effects on welfare aggregate (i.e., sum up) for each action? In that way, we calculate the action's overall effects on welfare. As already mentioned about the production of cheap meat, this calculation can be difficult since actions can have effects on many individuals at different times and places. Some of the effects may be very indirect or uncertain.
c. If the outcome of the action is uncertain, we need to assess for each state of nature (i.e., for each way the world may be if we perform the action in question), the value in terms of welfare of each outcome and the likelihood with which each of these outcomes will be realized. In that way, we perform an *expected-utility calculation*. For example, if I kill an animal, this action may be good or bad for the animal in question, depending on what would otherwise have happened to the animal, had I not killed it. I can never know for sure what would otherwise have happened, whether the animal would otherwise have had a good or bad future, but I can make an informed guess. Depending on what would otherwise have happened, killing the animal is good or bad for the animal. Effects on the welfare of others also need to be considered.
d. We need to know what the alternatives are and how each of them affects welfare. What the alternatives are often depends on who acts. For example, researchers may have different alternatives when it comes to animal experimentation than governments or consumers. A researcher who is opposed to an experiment may not have the option of forbidding the kind of experiment in question. She may only have the option not to perform it herself, in which case someone else may instead perform the experiment, perhaps even in a worse way. On the other hand, her refusal may inspire others and may ultimately contribute to a change of legislation. In contrast, a politician who opposes an experiment may have different options. If and only if a better alternative to the action is available, the action in question is wrong. So, it is not enough to assess whether the action has more positive than negative effects on welfare overall. Even if its effects are negative overall, it is the

'optimific' and therefore the right action just in case the alternatives are even worse. And even if an action's effects are good overall, it is the wrong action if there is at least one better alternative.

Of the animal experiments at hand, the first involves experiments on mice, the second on zebra fish and the third on monkeys. In the evaluation of these experiments, utilitarians take the effects of the experiments on these animals into account. After all, these are all sentient animals and (therefore) subjects of welfare. It matters to these animals what happens to them. Their lives can go better or worse for them. In principle, their welfare counts for just as much as everyone else's welfare, no matter what species the welfare subjects belong to. So, utilitarians need to assess, among other things, to what extent these animals' welfare is affected.

It seems clear that some actions that were performed during each of the three experiments were bad for the directly concerned animals and that there were better options available. For example, mice were kept in certain conditions and there are probably alternative conditions that allow mice to live better lives and some of the people involved in this experiment were probably in principle able to bring about these better conditions for these mice. For example, one can in principle take the mice out of the laboratory and either set them free (if it can be expected that they have better lives in that way) or house them somewhere else in better ways. Similarly, one could probably rescue the monkeys and allow the young to stay with their mothers. Everyone who caused one or more of these individuals to suffer or to experience less welfare than they would otherwise have could in principle have refused to do so. There is a question as to whether this would have been better for the animals in question. It would *not* have been better if someone else had then performed the very same actions on the very same animals. So, perhaps more than merely a refusal to experiment on them would be necessary to improve the lives of these animals. Probably an action that would have better served these animals was available to one or more people involved in the experiment.

Of course, the welfare of the involved animals is not the only thing that matters. Utilitarians need to assess the consequences for the welfare of *all concerned individuals*. Importantly, this includes the individuals that the agent could have benefited (or harmed) if she performed an alternative action. So, we need not only assess whether bringing the animals out of the lab would have been possible and whether this would have been better for them. We also need to explore how doing something else rather than performing the experiment would have been for others. How would it have affected the researcher, as well as her family, or colleagues? Would the very same experiment be performed on other animals instead? Would the liberation action receive media attention and inspire many people to change their minds about animal experimentation? Would the researcher be able to do other work that does more to maximize overall welfare? It is very difficult to say whether everyone involved in the experiment did the very best she could do in the circumstances. If a better action was available somewhere in the process, one or more agents acted wrongly either by experimenting on these animals in the first place, or in their performance of the experiment.

It is one question whether it was okay to participate in the experiment, and it is another question whether it was okay to allow it. Should, for example, governments or ethics committees allow that kind of experiment? Often this issue is decided based on a cost-benefit analysis. In contrast to cost-benefit analyses in other contexts, those that are typically performed to evaluate animal experimentation focus on consequences in terms of welfare rather than, for example, on economic effects. In particular, the focus in project evaluations ahead of animal experimentation is often on the potential societal benefits on the one hand and on the costs for the involved animals on the other hand. This looks like a utilitarian assessment. Nevertheless, cost-benefit analyses often fail to capture utilitarian considerations. They fall short in the following ways:

a. Narrow cost-benefit analyses typically do not take harm to animals as seriously as harm to humans. So, the utilitarian requirement of non-speciesism is often violated. If we would not, for the same benefits, allow the same amount of harm to humans that we inflict on the animals, we are speciesist and this is unjustified.
b. In narrow cost-benefit analyses, harm that is inflicted on the animals is typically rated as 'mild', 'medium' or 'severe', where painlessly killing the animals usually counts as 'mild' harm. Painlessly killing a human, in contrast, typically counts as inflicting severe harm. Indeed, if an experiment saves human lives, this counts heavily on the side of the benefits. As explained above, the dominant ethical theory about the harm of death, the deprivation view, implies that the harm of death consists in the deprivation of future life. So, ending a life at a particular time harms the animal, if (and only if) it would otherwise have had a good future. In many cases in which animals are killed during or after experiments, this deprives the animals of good futures that they could otherwise have had, if they were allowed to continue living in suitable conditions. Sometimes the animals are in such a miserable condition that continued life would be bad for them. In that case, killing them painlessly benefits them. (But then bringing them into such a miserable condition harmed them in the first place.) So, the harm of death for non-humans is often neglected in cost-benefit analyses of animal experiments.
c. In narrow cost-benefit analyses the range of alternative options is often not fully considered. It is a very important question what range of alternatives one should consider. The utilitarian answer is that this range should be as broad as possible. So, for example, with regard to the first experiment, the question should not be "Are there alternative ways of using CRISPR in order to produce albino mice?" Nor should the question be "Could I produce albino mice in alternative ways?" Instead, the question should be much broader. It should rather be: "Is there anything else I could do, other than an animal-based experiment, that would maximize the welfare in the universe or make lives better overall?" We should ask: "Among all uses of our time, money and resources, is this the one that most promotes welfare for all?"

So, a narrow cost-benefit analysis, as it is often performed in assessments of animal experiments, is unacceptable from a utilitarian perspective.

Doing such a narrow cost-benefit analysis for the cases at hand would be difficult enough. On the one hand, it would involve assessing the impact of the experiments on the welfare of the involved animals. So, one would need to ask how many animals were involved and to what extent their welfare was impacted. In that context, one would need to take breeding, the killing of these animals (including 'surplus' animals) and the housing conditions into account, which raises additional questions, as we saw above. On the other hand, one would need to assess the expected benefits of the experiment for human welfare. This is difficult since none of the experiments directly results in the elimination of any disease. Instead, at the very best, they help us gain some knowledge related to the disease. It is hard to say how useful this knowledge really is when it comes to reducing the disease-related harm. So, it is not enough to look at how serious the disease in question is and how many people are suffering from it. This is often done in narrow cost-benefit analysis. It always sounds impressive on the 'benefit' side that an experiment aims at fighting some serious and common disease. This, however, is not the information that we need. Instead, we need to know what exact *difference* the experiment in question really makes about disease-related harm. This is highly controversial, both in general and in specific cases. While performing a narrow cost-benefit analysis would therefore already be very difficult, I argue that it would also be useless from a utilitarian perspective, because so many crucial considerations would be excluded.

6.5 CONCLUSION

Thus, a utilitarian assessment of animal experimentation, which may at first sight seem easy and straightforward, turns out to be very complex in real-life cases. It is always possible to invent simple cases by way of thought experiments, where we just stipulate what the consequences of the various available actions are. In these cases, the utilitarian answer is clear: perform the action that has the best (expected) consequences. But in realistic cases it is difficult to oversee all (possible) consequences and to keep in mind all available alternatives. This is not a fault of the theory, which may still give us the correct answer as to what, in principle, we have most reason to do and why. It's just due to the complexity of real life and the limits to our knowledge and foresight that we often do not know what the action with the best consequences is.

For what it is worth, if I put on my utilitarian hat, I am not in favour of any of the three experiments presented in this volume. I think there are more effective ways of promoting welfare than performing that kind of experiments. I believe that improving public health, in particular mental health, and happiness, is a good way of improving overall welfare. But I think there are more effective ways of promoting welfare than performing these experiments. Much more can be done by way of disease prevention, such as promoting nutrition and lifestyle changes, as well as creating and maintaining healthy surroundings and relationships. We may never be able to prevent all suffering, but there is still a lot of low-hanging fruit that we can pick. I think it is more effective to do so, instead of (potentially) causing a lot of trouble in an effort of picking the fruit high up in the tree.

I think that, on the long run, adopting a caring relationship towards each other and towards non-human animals and our environment best helps us promote welfare. Sometimes that focus on empathy is contrasted with the utilitarian approach, since the latter seems to be rather cold-hearted and calculating. But this is wrong. A utilitarian approach to what we have reason to do is in fact maximally empathetic and caring. We are invited to empathize not only with those immediately in front of us, but with all individuals who are concerned by our actions, including those out of sight. That perspective of broad empathy might in some cases ultimately require making some lives worse to improve many others. But I think that there are so many options for win-win situations that allow us to promote welfare without causing harm. If we really start to care equally about all concerned individuals, including non-human animals, we will be more motivated to use the available win-win options and to find more of them. So, I think we do not have reason right now to turn to such drastic measures as described in the three experiments to maximize welfare.

Since the issues to be considered are so complex and wide-ranging, it is easy for utilitarians to disagree about the acceptability of the experiments discussed in this volume. Characteristic of the utilitarian stance is not a shared verdict about the acceptability of particular actions, but a common *perspective*. Utilitarians have in common that they take specific considerations to be relevant. They focus on assessments of consequences in terms of welfare. The utilitarian perspective excludes all considerations that do not ultimately come down to welfare. If someone claims that we shouldn't do this or that, the utilitarian asks: "Why not?" If some explanation is provided, the utilitarian asks: "Why does this matter?" Things, according to utilitarianism, ultimately matter if and only if they affect welfare in one way or the other. Animal experimentation, genetic modification or anything else that one can do is not in itself considered to be right or wrong. What one should do always depends on what would bring about most welfare (Figure 6.1).

REFERENCES

Bradley, B. 2009. *Well-Being and Death*. New York: Oxford University Press.

Crisp, R. 2006. *Welfare and the Good*. New York: Oxford University Press.

Lenman, J. 2000. Consequentialism and cluelessness. *Philosophy and Public Affairs* 29, 4: 342–370.

Sandoe, P., Corr, S., and C. Palmer. 2016. *Companion Animal Ethics*. Oxford: Wiley.

Shafer-Landau, R. 2012. *Fundamentals of Ethics*. New York: Oxford University Press.

Singer, P. 2011. *Practical Ethics*, 3rd ed. Cambridge: Cambridge University Press.

Sobel, D. 2007. The impotence of the demandingness objection. *Philosophers Imprint* 7, 8: 1–17.

Torrance, G. W., and D. Feeny. 1989. Utilities and quality-adjusted life years. *International Journal of Technology Assessment in Health Care* 5, 4: 559–575.

Višak, T. 2013. *Killing Happy Animals. Explorations in Utilitarian Ethics.* New York: Palgrave Macmillan.

Figure 6.1 Utilitarians can disagree on how to assess the potential welfare gains and losses related to a particular research project, but there is common agreement that the ideal is to maximize the overall welfare, which entails weighing the potential welfare loss of the animals up against the potential benefits. (Illustration Anna Kornum.)

7

Animal Rights View

SAMUEL CAMENZIND
University of Vienna

7.1 DEFINITION: ANIMAL RIGHTS VIEW

The presented position in this chapter is labelled *Animal Rights View* (ARV). It is an egalitarian sentientist position that claims that every sentient being should be granted three basic, individual rights – the right of freedom, right of defence and the right to assistance. Starting point is the broadly accepted premise that rights can be applied to every sentient animal and only to sentient animals, because they have an individual experienced welfare, which can be promoted or harmed (sentientism). Although it is an ongoing debate which animals qualify as sentient, according to empirical parameters sentience can be assumed in at least all vertebrate species (mammals, birds, reptiles, amphibians and fish) plus cephalopods and crustaceans (Low et al., 2012). Regarding the moral significance every individual sentient being has to be considered equally (egaliatrianism), although because of the wide variety of sentient beings, with species-specific and individual needs, the compliance with the three basic rights might lead to different

DOI: 10.1201/9780429428845-7

treatments in a specific situation. According to the ARV, basic rights guarantee that every sentient being, no matter what kind of species it is, has the possibility to live a good life of its own (rights theory). It is further important that the ARV presented here is concerned only with moral rights, which are independent of any political system and precede the implementation of rights in law and politics. Its relation to legal rights will not be further examined.

In the following the ARV and its political and philosophical background will be outlined first. Afterwards the three presented cases will be evaluated with the normative standards of ARV. It will be shown that the ARV would strongly limit the use of animals within scientific research.

7.2 BACKGROUND AND CONTENT

The ARV has to be understood as a close interrelation between philosophical, ethical and political doctrines. Historically and politically the ARV can be seen as an expansion of human rights as for instance mentioned in the *Universal Declaration of Human Rights* (1948), to sentient animals in general and not human animals only. Having their origin in the 18th and 19th centuries, the *Nonhuman Rights Project* (NhRP)[1] and the *Helsinki Declaration on the Rights of Cetaceans* (Low et al., 2012) are actual civil rights movements, which aim to embed pertinent rights in international and national laws to protect animals such as great apes, elephants, dolphins and whales from specific human use, such as hunting, human entertainment in zoos and circuses or harmful animal research.

Hence, ethically the ARV must be understood as an opposing position to human-centred (anthropocentric) positions and utilitarianism. Anthropocentric positions consider nonhuman animals only indirectly ethically relevant, e.g., because they are protected by a property law or because cruelty to animals will result in the delinquent habits in humans. Utilitarianism applies the idea of an impartial spectator, arguing that not only human but also animals' interests are morally relevant. However, due to the aggregation principle interests of a minority can be sacrificed to achieve the greatest good for the greatest number (e.g. Singer, 2011/1979: 20). This is something that proponents of the right view strongly oppose (cf. Regan, 2004/1983: Chapter 6.3).[2] An ideal society should grant its members maximal protection and not allow to trump the interests of a minority by the majority. In other words, all members of the moral community enjoy a strong protection within the ARV. This includes moral agents – beings who are morally responsible for their actions – and moral patients, beings who are protected by moral rights, but who are not able to follow moral rules, such as children, mentally disabled persons or animals.

The ARV further differs from the position discussed in the chapter on virtue ethics, because it focuses on action rather than the character traits of a moral agent and its starting point is neither the notion of "compassion" nor personal relationships as ethics of care does. But the rights view does not totally oppose the other

Henry Salt's *Animals' Rights Considered in Relation to Social Progress* (1894) and Leonard Nelson's *System of Ethics* (1932) may be viewed as antecedens of the animal rights view. Both emphasize the implementation of philosophical thoughts in to practice.

positions. It allows the weighing of interests within situations of moral dilemmas, and it allows compassion and relational duties among the members of the moral community as long as no right is violated. Further virtues play an important part to establish a stable society. In contrast to critical animal studies, whose perspective lies on institutional discrimination and (economical) power relations, the ARV focuses particularly on individuals and responsibilities between them.

The rights view that is presented here starts with other premises than ethics of care, virtue ethics and critical animal studies. It is based on three fundamental rights of individuals, which are:

- Right of defence, such as a right to life and bodily integrity.
- Right of freedom, such as a right of freedom of scientific research, individual lifestyle and freedom of movement and residence.
- Right to assistance, in a case of emergency or a right to education or pension.

To have a moral right is to have a moral claim *to* something and *against* someone. This means that each right must be linked with a corresponding duty, otherwise the claim cannot be implemented, because nobody would be responsible. Rights of defence correspond with a duty not to harm, rights of freedom with a duty not to interfere and right to assistance with a duty to assist. While every member of the moral community possesses the three basic rights, only moral agents have duties not to harm, not to interfere and to assist. Moral agents are beings who are capable of moral reasoning and who are responsible for their actions. Beings such as children, animals or people with mental disabilities, who are not able to be responsible for their action, are protected by rights, but don't have duties against the other members of the moral community.

The rights view assumes that the three basic rights are in the interest of every sentient being, although they prohibit some actions and restrict their freedom to a certain extent. Conceptually the right of defence is the strongest and precede the right of freedom and the right to assist. Otherwise, it wouldn't make any sense to claim the right of defence, if it would be generally overtrumped by the other two rights since then every action would be morally permitted. Hence, the right of defence trumps the other rights. The right of defence of all other sentient beings restricts my right of freedom and my duty to help. The only exceptions that permit to violate the right of defence are (i) cases of self-defence (and deadly force), (ii) when the rights holder gives her (informed) consent or (iii) if the rights holder benefits from the violation (e.g. in a case of emergency). In all other cases, the violation of right of defence is morally wrong and unpermitted.

What are the implications of the ARV for the three animal experiments? This question will be discussed in the next section.

7.3 CASE STUDIES: ANIMAL RIGHTS EVALUATION

Concerning all three cases an ethical evaluation has to ask first (Section 7.3.1) if the involved animals are sentient animals and therefore rights bearers. Secondly (Section 7.3.2), it has to be examined how the rights are affected and if any rights are violated. Thirdly (Section 7.3.3), if any rights are violated, it is necessary to evaluate if the violations can be qualified as exceptions (self-defence, consent, case of emergency). If no rights are violated or if the violated rights are qualified as exceptions, the animal research is morally justified in these cases. If rights are violated and they cannot be qualified as exceptions, then this type of animal research is morally wrong.

7.3.1 Are the research animals involved sentient creatures?

This question concerns the following animals: Regarding the mice case, the male mice (number unknown), the female "donor" mice (number unknown), surrogates and the 440 gene edited mice embryos lead to 174 grown mice. Regarding the zebrafish case it is morally relevant if the eggs and semen "donors" and CRISPR-edited zebrafish models are sentient. And in the rhesus macaque case, the question concerns the 32 female monkeys, semen donors (number unknown), 59 surrogate monkeys and the CRISPR-edited monkeys, which resulted in eight miscarriages, four full-term stillbirths and 14 live-born monkeys (nine of these were mutants).

Because mice, zebrafish and rhesus monkey all belong to the vertebrata, the scientific evidence that these animals possess a subjective welfare and therefore are sentient is given. In all three cases, the question remains difficult to answer: at what stage of development the CRISPR-edited animals can have sentient experiences? This is especially relevant for the monkey case, because if the miscarriages and stillbirths died before they are sentient, they will not qualify for moral consideration within the ARV. However, it is an ongoing debate at which stage of development vertebrates are assumed to be sentient. Some scholars draw the line at the moment the animals take their first breath, others suggest to draw it already after the last third of the prenatal development.

Unfortunately, data is missing of exactly how many animals are involved in the three cases. This means that the ethical evaluation lacks precision regarding the number of animals and their exact welfare state. This is not uncommon, because scientific papers are usually too short to allow detailed information about the welfare of research animals. Nevertheless, from an ethical perspective, this is not only regrettable but a serious shortcoming, because a comprehensive ethical analysis and evaluation is only possible to a certain point.

7.3.2 How are the rights of the involved sentient animals affected?

Although the specific handling of animals is not mentioned and detailed information about the exact procedures, interventions and the animals' welfare conditions are missing within short scientific publications, the descriptions are sufficient to recognize a large number of violations of animal rights – in fact, because the violations are legion, they will only be discussed exemplarily.

Regarding the setting of the experiments, it is relevant that most research animals have never lived in the wild. They are bred in captivity, serve human ends and die in captivity. In this context, their lives are strongly regulated according to human needs. Research animals cannot move freely, their circadian rhythm is adjusted to the human rhythm, they cannot choose or reject their mating partner (artificial insemination, embryo transfer) and for certain procedures they have to be fixed or anaesthetized. This means the right of freedom of probably all involved animals is violated.

Right of defence of female mice is violated in cases of invasive fertility treatments and their killing after mating to collect the embryos, and in stressful handling of surrogate mothers (e.g. surgery, embryo transfer, anaesthesia, postoperative discomfort). Regarding the knock-out mice, no information about possible harms is given. It is not possible to assess if any right of defence is violated.

The genetic editing of zebrafish involves stressful handling of donor fish (collecting eggs and semen) and it seems likely that zebrafish would experience periods of discomfort and lethargy due to hypoxia. In the cases of the CRISPR-edited zebrafish the affection of the right of defence is complex, because it is caused before the fish enters the moral community. But from the ARV it is wrong to breed animals that will likely have medical issues (e.g. brachycephalic syndrome within British bulldogs) or intentionally modify the genotype of an animal to achieve an animal model that expresses diseases.

The case of the induced muscle dystrophia in rhesus macaques, which involves invasive fertilization and obstetric procedures, shows a violation of the right of bodily integrity (right of defence). The acceptance of still births, miscarriages and difficult births puts a risk on the mother and therefore violates her right of defence. Further the separation of the mother and infants has a negative emotional effect on the infants and violates the right of freedom of the mother to care for her offspring. The risk that the increasingly reduced motor function may result eventually in paralysis and death clearly violates the right of defence of the CRISPR-modified monkeys.

To sum up, there is no doubt that the right of defence (right to life and bodily integrity) and the right to freedom are in all three cases violated in various (and not fully explored) ways. From the ARV these experiments are morally wrong if they cannot be qualified as exceptions, which has to be clarified in the next section.

7.3.3 Exceptions and other arguments to justify animal rights violations

As mentioned above (cf. 7.1), there are three exceptional cases which justify the violation of right of defence and right of freedom. These are (1) cases of self-defence (and deadly force), (2) when the rights holder gives her consent or (3) if the rights holder benefits from the violation (e.g. in a case of emergency). But none of these three exceptions applies to the three cases.

i. First of all, none of the three experiments can be qualified as self-defence. Neither the mice nor the zebrafish nor the rhesus monkeys are attacking the scientists, so they could defend themselves in a way that causes serious bodily injury or death to the animals. A grey area would be a pandemic situation, where humans and animals are forced to stay in quarantine (restrict their right of freedom) or are forced to be vaccinated (violation of bodily integrity), if the person or animal in question is a potential source of danger for others and if this is the ultimate measure, to contain a disease. Because neither the mice nor the zebrafish nor the rhesus monkeys are a potential source of danger for the human beings, the justification of self-defence is not applicable.
ii. Because neither the mice nor the zebrafish nor the monkeys are able to understand the complex research questions (some of them don't even exist before the experiment), the animals cannot consent to the experiment as a whole. Therefore, it is not possible to apply the consent criterion regarding the whole experiment. But cages, attachments and anaesthesia indicate, that animals don't participate freely and have to be forced to several, although not all, intermediate procedures.
iii. In the case of the CRISPR-edited mice, none of the mice benefit from the experiment. It belongs to basic research focusing on the development of efficient and affordable method to create knock-out animals as disease models. The CRISPR-edited zebrafish and rhesus macaque serve as an *in vivo* model, to simulate and study the heterotaxy syndrome or to study and cure Duchenne muscular dystrophia in humans. If they get cured during the experiment, one might say that they are beneficiaries. But given the circumstances that they were bred with the intention to express a disease and that they are likely to be killed after the research or used for other experiments, this positive assessment is invalid in an overall assessment. Further, none of the other involved animals (sperm and egg donors, surrogates) benefit from the experiments.

In the context of veterinary medicine, it is argued that animal research is morally permissible or even required, because it is also beneficial for many other animals of the same species. Proponents of the ARV reject this argument, because according to their position, a just society should reject to sacrifice a minority for the benefit of the majority. The function of the fundamental individual rights of humans and animals is exactly to prevent such cases.

The bottom line from an ARV is that no argument can be made to justify the violation of right of defence or the right of freedom in any of the three cases, (1) because they are not cases of self-defence (and deadly force), (2) because the rights holders don't give their consent and (3) because none of the rights holders benefits from the violation.

Regarding the human side of the experiment some proponents of animal research argue that scientists also possess a right to scientific freedom. Some argue further that scientists and doctors have a duty to assist humanity as such and especially to assist and cure their patients (Blumer, 2004). As mentioned in Section 7.1, it has to be stated against this argument that the right of freedom can never trump the right of defence. This means for example that the right to scientific freedom cannot overrule the right to bodily integrity or right to privacy. No psychologist is allowed to film someone during the night without her consent just because of his scientific interests. Nor is a food scientist permitted to poison someone to test a new product. Similarly, physicians are not allowed to kill one person to harvest organs to support his clients with them. The duty of doctors to assist simply stops when it implies the violation of a right of bodily integrity of others.

7.4 THE ANIMAL RIGHTS VIEW AND THE MORALITY OF BREEDING ANIMALS

What about the breeding of GMO animals or companion animals? Would these practices still be morally permitted within the ARV?

The similarities and differences between traditional breeding, genetic engineering and gene editing technologies are a complex matter and cannot be discussed comprehensively here. Nevertheless a few aspects will give insight into how the ARV will approach this issue. For the ARV the purpose of breeding is morally relevant and as the purposes animals are used for in science (e.g. basic and applied research, toxicology testing or gene pharming) are mostly not in the best interest of animals, they generally raise moral scruples for the ARV. Similarly, if companion animals are mainly bred for financial reasons as investment or business model, the rights view would morally condemn them. As the above-mentioned cases reveal the manipulation of the genome entails a potential health risk for the generated animals, it often involves the violation of the right of bodily integrity, during the process of harvesting oocytes or collecting donor cells. That's why they morally fall back compared to traditional breeding.

Whether breeding and keeping of companion animals are morally permitted in general is actually debated within the ARV. Some animal advocates condemn the domestication of animals and propose an "apartheid" policy that claims we should abolish all human uses of animals whatsoever. A second group recognize domestic animals and wild animals that live in the human domain as appreciated co-citizens and criticize only interactions that violate fundamental rights. In this case the companion animals or use of service or guarding dogs is not morally wrong per se (e.g. Donaldson and Kymlicka, 2011). But the legal property status of animals, the fact that we are allowed to buy and sell animals raise still more moral issues. The view presented in this chapter would probably accord to this second view. A third, intermediate position disapproves the new breeding of animals, but tolerates having animals from shelters or from finished animal research projects.

7.5 CONCLUSION AND OUTLOOK

What are the consequences of the animal rights position for animal research in general? Within the Animal Rights View, animal research is only permitted if the animals aren't harmed and no right of defence is negatively affected, such as

Figure 7.1 From an animal rights perspective the current practice of using research animals is very problematic. The goal is thus to end this practice as it violates the rights of the animals. A step in that direction is to focus the attention on one of the 3Rs: Replacement. One option is to, whenever possible, replace real animals with models of different kinds, in this case stuffed animals. (Photo Dorte Bratbo Sørensen.)

non-invasive behavioural or nutrition studies. All experiments that inflict pain or suffering or result in the death of animals are morally wrong. A permissible exception that may include a health risk for a treated animal is if the animal itself is ill and will benefit from testing a new therapy or medication. In this case the research is done in the best interest of the animal patient.

This also means that with regard to the 3Rs (Replacement, Reduction, Refinement) only the Replacement Principle would remain relevant to discuss. Because traditional animal research would be prohibited, a tremendous shift towards the development of alternative methods would be the case. Knowledge that could only be gained by harmful research has to be renounced. If these two factors – focus on alternative methods and waiver of some knowledge – would have a positive or negative impact for humanity compared to the actual situation remains an open empirical question. A realistic scenario has not been developed yet.

How far-reaching the consequences of the ARV are is still debated within the animal rights movement. In any case the ARV would have a huge impact on the society and our actual human-animal relationships that goes beyond animal research. Respecting fundamental, individual animal rights would strongly limit the use of animals in agriculture and entertainment such as zoos, circuses and hunting (Figure 7.1).

NOTES

1 Homepage of the NhRP: https://www.nonhumanrights.org/who-we-are/ [retrieved 14.11.2021].

2 Contemporary proponents of the animal rights view are Donaldson and Kymlicka (2011), Cochrane (2018) or Korsgaard (2018).

REFERENCES

Blumer, K. 2004. Ethische Aspekte zu Tierversuchen und das Solidaritätsprinzip. In: *Senatskommission für tierexperimentelle Forschung* (ed.): Tierversuche in der Forschung, Bonn: Lemmens Verlag- und Mediengesellschaft, pp. 27–29.

Cochrane, A. 2018. *Sentientist Politics. A theory of Global Inter-Species Justice.* Oxford: Oxford University Press.

Donaldson, S., and W. Kymlicka. 2011. *Zoopolis. A Political Theory of Animal Rights.* Oxford: Oxford University Press.

Nelson, N. 1956/1932]. *System of Ethics.* Trans. by N. Gutermann. New Haven: Yale University Press.

Korsgaard, C. M. 2018. *Fellow Creatures. Our Obligation to Other Animals.* Oxford: Oxford University Press.

Low, P., Panksepp, J., Reiss, D. et al. 2012. The Cambridge Declaration of Consciousness. Online: https://fcmconference.org/img/CambridgeDeclarationOnConsciousness.pdf [retrieved 19.5.2021].

Regan, T. 2004/1983. *The Case for Animal Rights.* Berkeley: University of California Press.

Salt, H. 1980/1892. *Animals' Rights. Considered in Relation to Social Progress.* Clarks Summit: Society for Animal Rights.

Singer, P. 2011/1979. *Practical Ethics.* Cambridge: Cambridge University Press.

8

The virtues and vices of germline editing research

CHERYL ABBATE
University of Nevada

8.1 VIRTUE ETHICS

Virtue ethics is a moral theory that is fundamentally concerned with the moral evaluation of the character of moral agents. Virtue ethics is thus primarily concerned with the following questions: *How should I live*? *What kind of person should I be*?

In answering these questions, virtue ethics tells us that we ought to do what the virtuous person would do and that we ought to live like how virtuous people live. And what's morally distinctive (and admirable) about virtuous people is that they possess a wide array of virtues, which are excellent character traits that are manifested in habitual action. Virtues essentially are traits that make people excellent *as people*.

To determine which character traits are excellent, we can, as a starting point, consider Aristotle's discussion of virtue and vice in *Nicomachean Ethics*. In this seminal text, Aristotle points out that there are certain spheres of human experience that all humans will inevitably encounter, and moral virtues are what dispose humans to choose and respond well in these spheres. On the other hand, the vices are what dispose humans to choose and respond defectively in these spheres.

DOI: 10.1201/9780429428845-8

As Aristotle pointed out, when it comes to (most of) the moral virtues, there are vices of both excess and deficiency. For instance, we, as humans, will all inevitably encounter situations that cause us to feel fear, and someone who habitually feels an *excess* amount of confidence in such situations (and thus someone who habitually and needlessly endangers herself) has a vice of rashness, while someone who habitually feels a deficient amount of confidence in such situations (and thus habitually flees in the face of any kind of danger) has the vice of cowardice. Those who habitually feel and act appropriately in such situations possess the virtue of courage. So, the courageous person neither flees instantly in the face of feelings of danger nor recklessly endangers herself; rather, she habitually feels the right (intermediary) amount of confidence in such situations. Moral virtue, then, as Aristotle puts it, is a mean between excess and deficiency.

Although there is no comprehensive list of virtues, Aristotle provided us with a starting point of 11 different spheres of human experience and their associated virtues and vices (Table 8.1).

Table 8.1 List of virtues and vices of excess and deficiency from the *Nicomachean Ethics* by Aristotle

Sphere of action or feeling	Vice of excess	Virtue	Vice of deficiency
Fear and confidence	Rashness	Courage	Cowardice
(Bodily) Pleasures and pains	Self-Indulgence	Temperance/ moderation	Insensibility
Giving and taking money (small sums)	Prodigality	Liberality	Meanness
Giving and taking money (major sums)	Tastelessness/ vulgarity	Magnificence	Pettiness
Honour and dishonour	Empty vanity	Proper pride	Undue humility
Anger	Irascibility	Good-tempered	Unirascibility
Speech (about the self)	Boastfulness	Truthfulness	Mock modesty
Conversation	Buffoonery	Ready-wit	Boorishness
General social conduct	Obsequiousness/ excessive flattery	Friendliness	Quarrelsomeness
Shame	Shyness/ bashfulness	Modesty	Shamelessness
Pain and pleasures about the fortunes of others	Envy	Righteous indignation/ proper judgement	Spitefulness

Aristotle's list, of course, is not complete. Another important moral virtue is compassion, which is the disposition to choose and respond well to the misfortunes of others. More specifically, it is a mean between the excess of overwhelming sympathy/softness and the deficiency of callousness. Another important virtue is justice, which is the disposition to give others what they deserve. Yet, as Aristotle plausibly argues, this virtue is unique in that it only has one extreme associated with it—injustice, which is the disposition to fail to give others what they deserve.

When we think of virtues, we often think about the moral virtues like compassion, justice, courage, and so forth. But there's also a second kind of virtue: intellectual virtue, which disposes humans to be good thinkers and learners. Examples include attentiveness, carefulness, curiosity, deliberative excellence, open-mindedness, prudence (practical wisdom), thoroughness, and so forth.

One thing that's distinct about virtue ethics is that it offers a contextual approach to morality. This means that, for a virtue ethicist, there isn't a straightforward answer to the question of whether something like lying is wrong. Whether a virtuous person would tell a lie depends on the particular context in which the lie occurs. Sometimes lying may be the compassionate thing to do, such as when one tells a lie to avoid hurting another's feelings. Other times, lying is done callously, such as when a lie is told in order to harm others.

8.2 VIRTUE ETHICS AND ANIMAL EXPERIMENTATION

Animal ethicists who are sympathetic to virtue ethics worry that utilitarianism suffers from a vice of excess and rights theory suffers from a vice of deficiency (Abbate, 2014; Merriam, 2012). As Garret Merriam (2012) notes, when it comes to animal experimentation, utilitarianism entails that it's justified to subject animals to excruciating suffering in an experiment that yields trivial benefits, so long as the benefits are enjoyed by a large enough number of people. For instance, severely hurting a few animals for the purpose of cosmetic testing could theoretically be justified on utilitarianism, so long as a sufficiently large enough number of people enjoyed the resulting product. On the other hand, rights theory seems to prohibit too much. After all, it seems to prohibit all forms of animal experimentation, even minimally invasive experiments that are necessary for preventing significant suffering and death.

Insofar as the central moral concern of virtue ethics is neither maximizing net happiness nor respecting the "rights" of individuals, virtue ethics may avoid some of the counterintuitive implications of rights theory and utilitarianism. Relatedly, given that virtue ethics rejects the notion that there are absolute moral principles and it endorses a contextual approach, it is flexible enough to satisfactorily address the complexities of morality. And a contextual approach to ethics seems especially useful when assessing the morality of various research studies, given that many present thorny moral dilemmas that are difficult to resolve in a satisfying way with inflexible utilitarian or rights-based principles.

When it comes to evaluating a particular animal experiment, instead of asking "what will maximize net utility?" or "will the moral rights of the individual be violated?" virtue ethicists will ask: What kind of person would perform such an experiment? In answering this question, a virtue ethicist will consider that (1) most animals used in animal research are sentient, which means that they have the capacity to suffer, and (2) most animal research causes animal suffering. And to fail to care about and be moved by an animal's capacity to suffer is to lack compassion. But a virtue ethicist will also note that, oftentimes, animal experimentation is conducted in the name of preventing human suffering, and to fail to care about and be moved by human suffering is also to lack compassion.

So, you might think that, for virtue ethics, animal experimentation just involves a conflict of moral virtues that can be settled by answering this question: Is compassion for animals or compassion for sick humans weightier in this experiment? Yet, the moral issues with animal experimentation run much deeper than this. In particular, considerations related to other moral virtues, such as truthfulness and justice, and the *intellectual virtues* play important roles in the virtue ethicist's moral assessment of animal research.

Consider, for instance, that animals are subjected to painful experiments usually in an attempt to derive some benefit for *humans*. And many people, researchers included, would find it appalling to harm humans in the same way we harm laboratory animals (for the same reasons). This is *speciesism*, which is an issue of justice (for more on speciesism, see Chapter 5). Moreover, using animals as human models also signifies an *intellectual* failing, insofar as testing on animals is usually an ineffective, and indeed counterproductive, way to promote human health (Akhtar, 2015; Engel, 2012). Some researchers, then, refuse to work with animal models altogether and publicly condemn animal research (Greek, 2010), which demonstrates the virtue of courage. And when scientists admit that animal research rarely yields knowledge that can be applied to humans, they demonstrate proper pride and truthfulness.

But it's unclear whether there currently are promising alternatives to using animals when it comes to germline editing research. Germline editing is the gene editing of only reproductive cells and it thus involves the alteration of the genes of zygotes or early embryos. Yet, this kind of research on humans is rarely funded and many countries have regulations that prevent human germline editing. For instance, in the United States, the National Institutes of Health (NIH) does not fund research that involves human embryo manipulation (Collins, 2015), and the Food and Drug Administration is prohibited from using funds to review research that involves the germline modification of human embryos (Cohen and Adashi, 2016). Such policies reflect the public's attitude towards human germline editing, which is one of unease (Funk et al., 2016). Indeed, when, in 2018, Chinese researcher He Jiankui claimed to have edited the genes of two human embryos that were brought to term, he was met with significant backlash from the public and scientists around the world (Hollingsworth and Yee, 2019).

Genome-editing research is also unique in that it likely will be used to benefit not only humans but also animals. For instance, it could prevent significant farmed animal suffering by preventing the killing of unwanted male baby chicks in the egg industry (Leenstra et al., 2011); it could be used to make animals more resistant to disease (Shriver and Mcconnachie, 2018); and it could be used to prevent high infant mortality rates, thereby decreasing wild animal suffering (Johannsen, 2017). So, even if the virtuous person is sceptical of traditional kinds of animal research, she may still feel positively about genome-editing research in animals. We'll now consider what a virtuous person might say in response to three germline editing studies that were recently conducted on animals.

8.3 VIRTUES, VICES, AND GENE-EDITED MICE AND MOSAICISM

Recently, researchers devised a way to use CRISPR/Cas9 to completely knock out the *Tyr* (Tyrosinase) gene in mice (at the zygote level), which caused albinism in 81 "knock-out" mice (Zuo et al., 2017). Although the albinism itself didn't have negative impacts on the welfare of the mice, this doesn't mean that the experiment is morally unproblematic. As noted, experiments like this one involve invasive fertility treatments, stressful handling, injections, surgery, and post-operative pain. While one might say that this just amounts to "mild" harm, a virtuous person would ask: if there were a study which involved subjecting human babies to non-therapeutic research studies that involved stressful handling, surgery, and post-operative pain, would we characterize the mistreatment of the babies as "mild"?

Moreover, animals permanently confined to research labs usually endure serious harms of deprivation, insofar as they are deprived of many kinds of positive experiences. Mice, for instance, have an interest in engaging in species-normal behaviour, such as play (Terranova and Laviola, 2005), but this is something they aren't always, if ever, permitted to enjoy when confined to research labs. Relatedly, laboratory cages are highly unnatural settings and thus prevent animals from behaving in natural ways, which arguably negatively impairs their welfare by, for instance, causing anxiety (Olsson and Sherwin, 2006). Indeed, research shows that providing mice with environmental enrichment, such as exercise tunnels, wheels, and chewable blocks, can help reduce the anxiety of artificial environments, such as laboratory cages (Olsson and Sherwin, 2006). And it's unclear whether Zuo et al. provided such stimulation for the mice.

The point here is that mice not only have the capacity to suffer, but they also have the capacity to experience joy and other positive emotions (Dolensek et al., 2020), and they are harmed when they are prevented from experiencing these positive emotions. So, to say that experiments that subject mice to only "mild" pain are morally unproblematic is to fail to appreciate their rich mental complexities. At the very least, it demonstrates an intellectual failing: the failure to acknowledge that when animals are permanently confined, they often endure serious harms of deprivation and felt anxiety, even if they don't suffer excruciating physical pain. At worst, it demonstrates a callous disregard for the welfare of the mice.

Of course, the deprivational harms and psychological suffering associated with confinement may be lessened in environments that are sufficiently enriched. Yet, animals used in research are normally subjected to artificial and restricted housing environments, such as small cages in windowless rooms filled with artificial lighting and noises (Akhtar, 2015). This is often due to space and funding restraints, and, in other cases, it is due to a blatant disregard for the well-being of the animals.

So, this experiment certainly caused harms to the animals involved. And given that Zuo et al. remain silent about the housing conditions of the mice, it's safe to assume that the mice used in this study, like most "lab animals," didn't enjoy sufficiently enriched environments and thus were also subject to deprivational harms. And a virtue ethicist will, at the very least, be moved by this. This doesn't, however, mean that she would refuse to partake in the research. After all, a virtuous person has compassion for *all* creatures who are vulnerable to harm. So, a virtuous person will be moved not only by the harm done to research subjects but also by the harm the study aims to prevent.

Arguably, a virtuous person will be disposed to cause a relatively minor harm when doing so is necessary to prevent a more significant harm. For instance, if a virtuous person finds herself in a situation where she can save someone's life only by pushing a second person out of the way, thereby causing a small harm to the second person, surely the virtuous person will choose to push the second out of the way. Refusing to do so because one has an overwhelming concern about the minor harm that may be done to the second would amount to excessive sympathy or softness for the second and a callous disregard for the one whose life is endangered. Arguably, a compassionate person is one who weighs harms appropriately and is disposed to prevent the most serious harms, even if this requires that smaller harms be done to others. Essentially, a virtuous person will draw a moral distinction between (1) needless harm and (2) necessary harm, that is, harm that is necessary for preventing a more significant harm. The virtuous person's sense of compassion will move her to prevent significant harm and, in tragic, forced-choice situations, her moral courage will move her to cause necessary harm in the name of preventing more significant harm.

To understand whether, and to what degree, the Zuo et al. research has the potential to reduce harm to others (humans and animals), we need to first have a basic understanding of *mosaicism*, which, until recently, was prevalent in the majority of animals who were edited with CRISPR. When mosaicism occurs, editing occurs only in a portion of cells. This means that when a mosaic mutation occurs during the use of CRISPR, an embryo ends up with a mix of edited and unedited cells after a gene is edited. So, if CRISPR is used in an attempt to prevent some genetic disease and the end result is that some of the cells in the embryo do not have the desired edit, this disease may still be present. Overcoming mosaic mutations when using CRISPR is thus important for the success of CRISPR technology.

CRISPR has the potential to make genome editing more simple, efficient, rapid, accurate, easier, and cheaper in comparison to older gene-editing technologies, and thus it's commonly referred to as a "revolutionary gene-editing system" (Tabb et al., 2021). And because mosaicism is a major obstacle to this promising technology (Mehravar et al., 2019), research that seeks to overcome the problem of mosaicism for CRISPR would be deemed worth pursuing by virtue ethicists. But, of course, this doesn't mean that any method of attempting to overcome mosaicism would be endorsed by a virtue ethicist, as some methods may involve needless suffering.

Zuo et al. propose a new strategy for overcoming the problem of mosaicism for CRISPR, and they suggest that their strategy will make genome editing through CRISPR more efficient. In support of this claim, they point out that they used a modified C-CRISPR approach, and, in a single step, completely knock out the *Tyr* gene in mice. But one thing to note is that strategies for reducing or eliminating mosaicism may depend on species, which means that an approach that works on reducing mosaicism when gene-editing species X may not work on species Y (Mehravar et al., 2019). This means that even if mosaicism was overcome in the mice in this study, this doesn't necessarily tell us anything useful when it comes to genome editing in other species. And, as a virtue ethicist will point out, honest researchers would emphasize that their results cannot be extrapolated to other species, so as not to mislead the public into thinking their research accomplished more than it actually did.

Relatedly, a virtue ethicist will wonder why the researchers focused on mice and whether there is some special importance in furthering genome-editing research on mice. Some scientists think that genome editing on mice is important because mice are often used as models for human disease. Yet, there is significant evolutional distance between mice and humans, so a virtue ethicist will question the widespread practice of using mice as models for human disease. For instance, mouse and human genomes are only about 85 percent the same, and many DNA variations and gene expression patterns aren't shared between humans and mice (Lin et al., 2014). Indeed, the genomic responses to different acute inflammatory stresses in humans are not reproduced in mouse models, leading some researchers to conclude that the mouse models normally used to study human inflammatory diseases have been misleading (Seok et al., 2013). Other researchers note that humans have a type of cell in their cerebellum that doesn't exist in mice, which makes them poor models for human brain research (Haldipur et al., 2019). Consequently, researchers are becomingly increasingly worried that using mice models has led to wasting years and billions of dollars chasing false leads (Kolata, 2013). Researchers who waste funds in this way seem to have the vice of what Aristotle called meanness, insofar as they seemingly readily take much money (e.g., taxpayers' money that is given to them through public funding), but fall short in giving back "wealth" (e.g., helpful medical advances) to the public.

Interestingly, monkeys were used in a different part of this study, and, in the abstract of the paper, the researchers emphasize that a virtue of their approach is that it will help in "rapidly establishing gene-edited monkey models" (Zuo et al., 2017, p. 933). Yet, they remain silent about whether a virtue of their approach is that it will "rapidly establishing gene-edited *mice* models." And this silence about the value of creating knock-out mice suggests that the researchers really didn't have a good reason for using mice in the study. This, then, indicates that the researchers lack prudence (or practical wisdom), which is the intellectual virtue that enables people to reason rightly about practical matters. It is the virtue of thought that would, for instance, enable researchers to calculate the right means to the end of human health.

Finally, and perhaps most importantly, it should be noted that strategies for overcoming mosaicism when using CRISPR in mice are already available (Chapman et al., 2015; Jacobi et al., 2017; Quadros et al., 2017; Wu et al., 2015). Thus, a virtue ethicist will question whether Zuo et al.'s research on mice really provides us with practically useful information. While the researchers claim to provide a new method for overcoming mosaicism, "new" doesn't mean "better" or "relevantly different." Consider, for instance, that while producing genetically modified animals without mosaicism in a single step is a procedure that's apparently useful for large animals, it's unclear whether the "single-step" method is beneficial when it comes to smaller animals, such as mice. So, again, it seems that using mice in this study was unnecessary, and the fact that the researchers experimented on them nevertheless demonstrates both meanness and a lack of compassion for the mice. And the fact that these scientists claim to have done something that brings them glory (i.e., something "new") when they arguably haven't actually done so indicates that they have the vice of boastfulness.

8.4 VIRTUES, VICES, GENE-EDITED ZEBRAFISH, AND HETEROTAXY

In the Perles et al. (2015) study, CRISPR was used on zebrafish to enhance understanding of complex genetic causality involved in human heterotaxy. In the study, the mmp21 gene of zebrafish embryos, which is the equivalent of the human MMP21 gene, was deleted, and, as a consequence, this resulted in cardiac looping defects in 27 of the 113 CRISPR-treated zebrafish. The researchers thus claimed to have established a causal link between mutations in the MMP21 gene and phenotypic expression of heterotaxia.

Like mice, zebrafish are sentient creatures and have emotionally rich lives. They, for instance, experience fear and anxiety (Jesuthasan, 2012). The handling, egg and semen collection, and cardiac looping likely caused, at the very least, discomfort in the zebrafish, and it's unclear whether they were given anaesthesia. It's also unclear how the researchers determined which fish developed cardiac looping disorder. Were the fish killed to determine this? And how were they killed?

It's also important to note that this study is aimed specifically at benefiting *humans*, so a virtue ethicist will rightly wonder why zebrafish were used, when they are so different from humans. In fact, zebrafish are even more different from humans than are mice, given that only approximately 70 percent of human genes have at least one obvious zebrafish orthologue (Howe et al., 2013). Although the relevant human gene in this study has a zebrafish orthologue, we must remember that there is often variation in gene expression from species to species. Thus, as Akhtar (2015) notes, "consistent phenotypes (observable physical or biochemical characteristics) are rarely obtained by modification of the same gene, even among different strains of mice." This suggests that the observed cardiac looping effect that occurred in the "knock-out" zebrafish doesn't give us good reason to believe this same effect will occur in humans with a deleted MMP21 gene.

It's worth noting that scientists often conduct research on mice and zebrafish not because they are ideal models for humans but because it's easier. For one, mice and zebrafish are cheap and breed quickly. And in some countries, there are fewer regulations that govern the use of mice and fish in research, which provides researchers with more flexibility. For instance, in the United States, the Animal Welfare Act (AWA) doesn't apply to mice and fish, and thus researchers aren't required by the AWA to provide mice and fish with pain relief, which in turn reduces research costs. Relatedly, in many countries, it's easier to get approval to research on mice and zebrafish than it is to get approval to research on species that are more genetically similar to humans, such as monkeys. And scientists certainly are subject to less public backlash when they test on mice and zebrafish as opposed to when they test on monkeys. A virtuous person, then, will wonder whether, by choosing to experiment on mice and zebrafish, the researchers in the first and second studies were more concerned with generating a quick publication, and less concerned with making meaningful advancements for human health, which would surely demonstrate the vice of meanness. Relatedly, a virtuous person will wonder whether these researchers were cowardly, insofar as by experimenting on mice and zebrafish, they risked little public backlash, given that the public tends to be less concerned about these species of animals.

Relatedly, a virtue ethicist will want to know: Can this research be applied in a practical way, moving forward, in an attempt to help humans? Would it really be prudent to start editing the *human* MMP21 gene just because there is a causal link between mutations in the *zebrafish* mmp21 and phenotypic expression of heterotaxia? If the answer is no, a virtue ethicist will wonder if the researchers were wasteful—and wasteful of highly valuable (and limited) research resources.

And even if it's prudent to use zebrafish in research about the human MMP21 gene, we still must ask: Is the research itself something a virtuous person would endorse? As noted earlier, editing human germline cells is widely considered an unacceptable scientific practice, and it's unclear when, if ever, it will become normalized in science. So, even if it's prudent to edit the *human* MMP21 gene in light of this study's results, it's unclear when scientists will be able to do so through germline editing. For all we know, by the time human germline editing becomes a normalized scientific practice, significant scientific advancements regarding human heterotaxy will have been made, rendering the results from this research useless. So, a virtue ethicist likely would say that conducting this research right now, when editing human germline cells is practically non-existent, demonstrates the intellectual vice of imprudence as well as moral vices, such as impatience and, as noted, meanness.

8.5 VIRTUES, VICES, GENE-EDITED MONKEYS, AND DUCHENNE'S MUSCLE DYSTROPHY

In this study, CRISPR was used to target the dystrophin gene in monkeys to create mutations that lead to Duchenne muscular dystrophy (DMD) in 14 live monkeys. The main goal of the study was to determine whether CRISPR can efficiently generate monkey models of human diseases. Because the researchers found, through tissue sampling, that there was frequent disruption of the normal muscle structure in the monkeys who were subject to mediated mutations that resemble the mutations in human DMD patients, the researchers concluded that "CRISPR/Cas9 can efficiently generate monkey models of human diseases" (Chen et al., 2015, p. 3764).

To begin with, this conclusion overstates the study's results. At best, the researchers could conclude that "CRISPR can efficiently generate monkey models of *DMD*, regardless of inheritance patterns." Yet, even this conclusion overstates the research findings and indicates a failure of deliberative excellence. After all, at the time of the study, none of the gene-edited monkeys had developed obvious behavioural or movement abnormalities that are characteristic of human DMD. And because, at the time of publishing the study, the 14 live Cas9-targeted monkeys were younger than 6 months, the researchers admit that they didn't know whether the monkeys will show any phenotypes as they develop into adulthood.

It's worth noting that the researchers claim that mice are poor models for DMD research because mice do not express human-like symptoms (for example, mice express much milder phenotypes). So, if the researchers agree that mice aren't good models of DMD because they don't express human-like symptoms and phenotypes, it's irrational for them to conclude that their gene-edited monkeys are good models of DMD before determining whether they will express human-like symptoms and phenotypes.

A virtuous person will then wonder whether the researchers, by publishing their results before the gene-edited monkeys reached adulthood, were in a rush to publish, prioritizing a quick publication over a thorough study, which demonstrates, at the very least, a deficient amount of carefulness and thoroughness. Relatedly, the researchers' assertion that "CRISPR/Cas9 can efficiently generate monkey models of human diseases" with insufficient supporting data demonstrates the vices of vainness ("empty vanity") and boastfulness, as they essentially make excess claims about the merits of their research.

8.6 SHARED MORAL AND INTELLECTUAL VICES

All three studies failed to report what happened to the animals after the experiments ended. The unfortunate reality is that most laboratory animals are either killed after research or "re-used" in subsequent research. Yet, both options are seriously problematic from a virtue perspective. Indeed, both amount to viewing sentient animals as disposable and/or as mere research tools and fail to show basic compassion for the animals. Arguably, justice demands that some form of restitution be given to the animals who are harmed in scientific research and that they not just be disposed of or "repurposed" after the research ends. Relatedly, if animal research is sometimes a "necessary evil," justice arguably requires that the burdens be spread about more equally and not repeatedly placed on the same animals.

On a separate but related note, we must remember that a virtuous researcher not only devotes herself to research that has the potential to reduce vast suffering, but she also communicates honestly, clearly, and candidly with the public about her research (Figure 8.1). Thus, a virtue ethicist will wonder why the researchers, in their publications, neither reported what happened to the animals after the experiments ended nor detailed the ways in which the animals were housed and handled. Given that scientists are quick to report on the *potential* benefits of their research, it's curious that they remain silent about the *actual* harms caused by their research. This demonstrates a vice of dishonesty, and in particular, a vice of selective reporting. It moreover, in some sense, demonstrates the vice of shamefulness, insofar as the scientists fail to confront the harms they are regrettably responsible for, perhaps out of fear of

Figure 8.1 Being a virtuous researcher takes not only scientific knowledge but also an understanding of the ethical issues at stake. Thus, training of people involved with laboratory animals should include ethics as part of the education. (Illustration Anna Kornum.)

dishonour. And, ironically, at the same time, the scientists also demonstrate the vice of shamelessness, in that they openly and seemingly without regret admit to exploiting innocent and defenceless animals.

This vice of selective reporting and, relatedly, grandiose reporting is a pervasive problem in animal research. As I've noted, the researchers in the case studies weren't entirely forthcoming about the value of their studies, insofar as they overstated the usefulness of their research and/or excluded relevant information about the limitations of the practical applications of their studies. The presentation of the results was, at best, careless, and, at worst, dishonest. A virtue ethicist will thus not only call for more compassion for the animals used in research and less waste of valuable resources, but they will also enjoin researchers to be more truthful and humble in their reports and encourage them to make better use of their practical reasoning skills, so that they calculate correctly towards the end of human health (i.e., pursuing the right means to the end of human health) and not just the self-serving end of research productivity.

REFERENCES

Abbate, C. 2014. Virtue ethics and animals: A minimally decent ethic for practical living in a non-ideal world. *Journal of Agricultural and Environmental Ethics* 27: 909–929.

Akhtar, A. 2015. The flaws and human harms of animal experimentation. *Cambridge Quarterly of Healthcare Ethics* 24: 407–419.

Chapman, K. M., Medrano, G. A., Jaichander, P., et al. 2015. Targeted germline modifications in rats using CRISPR/Cas9 and spermatogonial stem cells. *Cell Reports* 10: 1828–1835.

Chen, Y., Yinghui, Z., Yu, K., et al. 2015. Functional disruption of the dystrophin gene in rhesus monkey using CRISPR/Cas9. *Human Molecular Genetics* 24: 3764–3774.

Cohen, G., and E. Adashi. 2016. The FDA is prohibited from going germline. *Science* 5: 545–546.

Collins, F. 2015. Statement on NIH funding of research using gene-editing technologies in human embryos. The National Institutes of Health. Retrieved from: https://www.nih.gov/about-nih/who-we-are/nih-director/statements/statement-nih-funding-research-using-gene-editing-technologies-human-embryos (accessed August 31, 2021).

Dolensek, N., D. Gehrlach, A, Klein, and N. Gogolla. 2020. Facial expressions of emotion states and their neuronal correlates in mice. *Science* 368: 89–94.

Engel, M. 2012. The commonsense case against animal experimentation. In *The Ethics of Animal Research*, ed. J. Garrett, 215–236. Cambridge, MA: Massachusetts Institute of Technology.

Funk, C., Kennedy, B., and E. Podrebarac Sciupac. 2016. U.S. public opinion on the future use of gene editing. Pew Research Center. Retrieved from: https://www.pewresearch.org/science/2016/07/26/u-s-public-opinion-on-the-future-use-of-gene-editing/ (accessed August 31, 2021).

Greek, R. 2010. Medical research with animals. In *Animal Rights and Animal Welfare*, ed. M. Bekoff, 373–377. Westport, CT: Greenwood Press.

Haldipur, P., Aldinger, K., Bernardo, S., et al. 2019. Spatiotemporal expansion of primary progenitor zones in the developing human cerebellum. *Science* 25: 454–460.

Hollingsworth, J., and I. Yee. 2019. Chinese scientist who edited genes of twin babies is jailed for 3 years. CNN. Retrieved from: https://www.cnn.com/2019/12/30/china/gene-scientist-china-intl-hnk/index.html (accessed March 8, 2022).

Howe, K., Clark, M. D., Torroja, C. F., et al. 2013. The zebrafish reference genome sequence and its relationship to the human genome. *Nature* 496: 498–503.

Jacobi, A. M., Rettig, G. R., Turk, R., et al. 2017. Simplified CRISPR tools for efficient genome editing and streamlined protocols for their delivery into mammalian cells and mouse zygotes. *Methods* 121: 16–28.

Jesuthasan, S. 2012. Fear, anxiety, and control in the zebrafish. *Developmental Neurobiology* 72, 3: 395–403.

Johannsen, K. 2017. Animal rights and the problem of r-strategists. *Ethical Theory and Moral Practice* 20: 333–345.

Kolata, G. 2013. Mice fall short as test subjects for some of humans' deadly ills. *The New York Times*. Retrieved from: https://www.nytimes.com/2013/02/12/science/testing-of-some-deadly-diseases-on-mice-mislead-report-says.html (accessed August 31, 2021).

Leenstra, F., Munnichs, G., Beekman, V., et al. 2011. Killing day-old chicks? Public opinion regarding potential alternatives. *Animal Welfare* 20: 37–45.

Lin, S., Lin, Y., Nery, J., et al. 2014. Comparison of the transcriptional landscapes between human and mouse tissues. *Proceedings of the National Academy of Sciences* 111: 17224–17229.

Mehravar, M., Shirazi, A., Nazari, M., and M. Banan. 2019. Mosaicism in CRISPR/Cas9-mediated genome editing. *Developmental Biology* 445: 156–162.

Merriam, G. 2012. Virtue, vice, and vivisection. In *The Ethics of Animal Research*, ed. J. Garrett, 125–146. Cambridge, MA: Massachusetts Institute of Technology.

Olsson, I., and C. Sherwin. 2006. Behaviour of laboratory mice in different housing conditions when allowed to self-administer an anxiolytic. *Laboratory Animals* 40: 392–399.

Perles, Z., Moon, S., Ta-Shma, A., et al. 2015. A human laterality disorder caused by a homozygous deleterious mutation in *MMP21*. *Journal of Medical Genetics* 52: 840–847.

Quadros, R.M., Miura, H., Harms, D.W., et al. 2017. Easi-CRISPR: A robust method for one-step generation of mice carrying conditional and insertion alleles using long ssDNA donors and CRISPR ribonucleoproteins *Genome Biology* 18: 92–106.

Seok, J., Warren, H., Cuenca, A., et al. 2013. Genomic responses in mouse models poorly mimic human inflammatory diseases. *Proceedings of the National Academy of Sciences* 110: 3507–3512.

Shriver, A., and E. Mcconnachie. 2018. Genetically modifying livestock for improved welfare: A path forward. *Journal of Agricultural and Environmental Ethics* 31: 161–180.

Tabb, M., Gawrylewski, A., and J. DelViscio. 2021. What is CRISPR, and why is it so important? *Scientific American*. Retrieved from: https://www.scientificamerican.com/video/what-is-crispr-and-why-is-it-so-important/ (accessed August 31, 2021).

Terranova, M., and G. Laviola. 2005. Scoring of social interactions and play in mice during adolescence. *Current Protocols in Toxicology* 26: 13–10.

Wu, Y., Zhou, H., Fan, X., et al. 2015. Correction of a genetic disease by CRISPR-Cas9-mediated gene editing in mouse spermatogonial stem cells. *Cell Research* 25: 67–79.

Zuo, E., Cai, Y., Li, K., et al. 2017. One-step generation of complete gene knockout mice and monkeys by CRISPR/Cas9-mediated gene editing with multiple sgRNAs. *Cell Research* 27: 933–945.

9

An ethic of care critique of induced genetic mutation in animals

JOSEPHINE DONOVAN
University of Maine

9.1 ETHIC OF CARE

The feminist ethic of care is a dialogical ethic that requires paying attention to an animal's own feelings about his or her treatment. It requires listening to his or her voiced or otherwise expressed point of view (Donovan, 2006, 2017). As an ethic of loving protectiveness, the feminist ethic of care derives from women's time-honoured and nearly universal role in human society of caring for children and others for whom one is responsible (see Ruddick, 1980). Care involves attending to and respecting the needs of those within one's sphere of influence and awareness, recognizing that they are independent entities with needs and interests apart from one's own, respecting their dignity, and trying to enable them to live optimally in accordance with their needs and wishes, while acknowledging their vulnerability.

Such an approach when applied to animals requires a careful focus on the animal's reaction or, when that is not possible, use of the empathetic moral imagination to appreciate what that animal's reaction is likely to be. Knowledge of an animal's generic identity and personal history is useful, but in general, as we share

DOI: 10.1201/9780429428845-9

communicative faculties with all animate creatures, it is not difficult to ascertain their wishes: to wit, that they do not wish to be treated in painful, stressful, and otherwise demeaning and disruptive ways. Nor do they wish to be killed or to have their basic genetic identity disenhanced. Even the most primitive of animal creatures—one-celled eukaryotes—know what is good or bad for them and express by movement their understanding of these realities—by moving towards what is beneficial and away from what is harmful. Humans have known about animals' feelings for millennia but have chosen for the most part to ignore this knowledge on the theory that animals are inferior to humans and that humans can therefore use them for their own purposes, regardless of how such use may impact the animal's own life experience.

Based on our knowledge of animals' wishes we can formulate, from an ethic of care position, our own ethical response, which is to respect their wishes (when such wishes do not immediately endanger one's own life or the lives and welfare of those for whom one is responsible). One way to estimate what an animal's reaction to any of the experiments described in this book would be to imagine asking for his consent—which would be required of a human subject in any lab experiment. Were any of the creatures involved in these experiments—mice, zebrafish, monkeys—asked their consent—after its procedures and purposes were carefully explained to them—it is quite clear that their response would be a resounding "No." They would be running or swimming away from the experimenter as fast as they could.

A certain humility is required in the caring praxis, a recognition that life is a complex and interdependent matrix of relationships, a complete understanding of which is beyond human ken. Interfering with one aspect of the matrix may have unintended consequences elsewhere in the network. Caring, therefore, requires a respect for the complex design that has evolved in all natural creatures, who are born with an inherited potential (*entelechy*) and who strive in their lives to realize what is good for them according to that design or *telos*. A caring ethic therefore requires respecting the needs and wishes of creatures to realize their own good.

Such an attitude counters the historic Western, masculine domination of the natural world, which is conceived—at least since the early modern period—instrumentally, as fodder for human purposes, with little or no care given to the ethical significance of other creatures' own wishes and purposes. By contrast, an ethic of care maintains that humans' relationships with animals should be dialogical. Humans should listen to what animals are communicating about how they wish to be treated and their views or standpoints should be respected and incorporated into any human decisions about their treatment and reflected in that treatment.

An attitude of humility recognizes, moreover, that nature, and in particular natural genetic processing, is an infinitely complex phenomenon. Like all of nature, genes are not predictably mechanical entities; they are pleiotropic, that is, they have multiple and often unpredictable effects; they control "for multiple, often unrelated phenotypic traits" (Schultz-Bergen, 2018, p. 229). This means that in gene-editing, even with the new relatively precise CRISPR technique, "we may produce an unintended phenotype change along with our intended one ... Hence the phenomenon of unintended gene expression suggests that there will always be a risk of unintended gene expression" (Schultz-Bergen, 2018, p. 230). Compounding the problem of accurate genetic engineering are the so-called epigenetic factors, "chemical compounds that are external to the genome but... can control when and how genes get expressed." These may interfere with and distort gene-editing so that "off-target edits" may occur (Schultz-Bergen, 2018, p. 228).

As living animals are the victims of these experiments, they are the ones who experience the mutilated bodies that result when genetic engineering goes awry. It is, in short, often a "phenomenological nightmare for the animals involved" (Weisberg, 2015, p. 45). Moreover, of course, these failed experiments—"collateral damage"—are usually killed.

Philosopher Mary Midgley condemned as an "exuberant power fantasy" (2000, p. 13) such projects, deploring the hubris of biotechnology for its instrumentalist mechanistic view of nature, wherein "a colossally complex system with its own laws" is turned into "a consignment of inert raw material laid out for our use" (2000, p. 12). Jennifer Doudna, the Nobel-Prize-winning biochemist who invented the CRISPR technique, acknowledged that it now gives humans "the power to radically and irreversibly alter the biosphere," enabling us "to bend nature to our will." Not surprisingly perhaps, Doudna herself worried that like Frankenstein she may in the CRISPR technique have "created a monster" (Doudna and Sternberg, 2017, p. 119, p. 117, p. 200).

9.2 GENE-EDITING MICE TO INDUCE ALBINISM

Were one to ask the original mother mice in the experiment designed to induce albinism whether they wished to have needles injected into their stomachs repeatedly, to be pumped full of fertility-promoting hormones, which increase anxiety and emotional distress, and then to be killed, so that their embryos may be "harvested" for the gene-editing procedure, the answer, we know, would be "No." This, according to an ethic of care, is important ethical knowledge that we humans should not ignore or override.

In this experiment the CRISPR technique was then applied in vitro to 440 harvested embryos with the Tyr gene, which effects the production of melanin for pigmentation, disabled or "knocked out." The embryos were then implanted in surrogate mothers. 174 thus genetically mutated mice were born as complete or partial albinos, that is, colourless or without pigmentation. It is not clear what happened to the other 260 embryos or baby mice. If these were miscarriages, or if the embryo failed to implant, there was likely accompanying pain and distress for the surrogate mothers. If they were born as defective baby mice, they were undoubtedly killed.

As for the remaining partial or complete albinos, it should be noted that albinism is not just an aesthetic matter of colour or lack thereof; it often involves vision impairment. How many of these mice were born blind or visually impaired as a result of the experiment is unclear, but presumably many may have been, an ethically deplorable result which no creature would choose for herself.

To what extent the colour change affected the social interactions among the mice is also unclear, if the albino mouse was housed with brown ones. An ethic of care holds that individual creatures cannot be understood apart from their cultural and physical environment—their *Umwelt*, as characterized by Jakob von Uexküll (2010). Brown mice evolved with that colouration because it enabled them to survive in a particular habitat and with particular cohabitants. A white mouse would be robbed of that protection and thus more vulnerable in a natural setting. Granted, the laboratory cage is not a natural setting, but destroying a mouse's natural capability divorces him even further from his natural ecological niche.

9.3 GENE-EDITING ZEBRAFISH TO REARRANGE THEIR INTERNAL ORGANS (HETEROTAXY)

Fish have feelings, as Jonathan Balcombe has revealed in an important book on the subject (2016). Indeed, "Zebrafish can get depressed and respond to the same antidepressant drugs humans do" (Montgomery, 2019, p. 16). The zebrafish in this experiment, designed with their internal organs rearranged, are thus not objects; they are individual subjects with feelings, wishes, and needs, which, according to an ethic of care, we humans are obliged to honour.

To cause these living creatures to be born with severe birth defects that will affect their ability to function normally, in accordance with their *telos*, is thus morally objectionable. Between 20% and 40% of the fish thus genetically distorted in this experiment were born with cardiac looping defects, a malformation of the heart that causes deficient oxygenation. These fish thus would have been impaired in their daily activities due to a lack of energy.

Since the deletion of the mmp21 gene is thought to induce heterotaxy, that is, abnormal rearrangement of internal visceral organs, the fish so treated would likely have had other serious malformations, such as to the lungs, spleen, liver, and intestines. Needless to say, the malformations of any of these organs would likely cause pain and distress to the creature and premature death. Thus impaired, the animal would struggle to carry on her normal behaviour but would be prevented in doing so by the genetically induced defects. An animal thus "disenhanced," that is, deprived of normally occurring faculties or coerced to endure artificially imposed impediments, is robbed of the capacity to strive for and realize her natural good. Such coerced deprivation is ethically objectionable because it elides and vitiates the animal's own needs and wishes.

Moreover, it robs the animal of dignity. It may seem anthropomorphic to consider that a fish could have dignity. But dignity at its simplest means being able to carry on independently in effecting the realization of one's own individual and species-specific life purposes, or, at a minimum, being entitled to respect for that capability.

A poignant article by Suzanne Laba Cataldi (2002) describes how she came to realize how the concept of dignity applies to animals when she observed some bears performing in a circus, riding bicycles or dressed in silly costumes, etc. Such coerced behaviour, she felt, seemed to violate the animals' dignity: "what is so painful about looking at these bears... is their lack of a... certain bear dignity" (Cataldi, 2002, p. 107). It was a matter, she realized, of not letting the bears be who they naturally are—that is, *bears*, not ornaments or toys for human amusement. Dignity, she decided, means being "self-possessed" and in control of one's own life (Cataldi, 2002, p. 115)—in the case of animals, pursuing one's own species-specific *telos*.

In the case of the zebrafish: having one's internal organs impaired so as to be unable to function as a fish—to swim freely, to pursue food, to engage in one's natural behaviour, and to control one's movement—means being deprived of the possibility of living one's life as a fish-being, instead of as an experimental object designed for human purposes. Such distortion of a fish's natural being robs it of dignity.

9.4 GENE-EDITING TO INDUCE MUSCULAR DYSTROPHY IN RHESUS MONKEYS

The monkeys in the experiment designed to induce muscular dystrophy are no longer seen as living subjects with their own needs and wishes; rather, they are "models," that is, scientifically designed simulacra stripped of feelings and thoughts—in other words, phenomenologically disensouled, of interest only as physiochemical mechanisms. Were these animals considered living subjects and their feelings and wishes considered of ethical significance, these experiments would (and should) never have occurred. For, without question, were these animals asked if they would willingly undergo the pains, stresses, and agonies of disablement, their answer would be "No."

As with the mice, the mothers and fathers of the monkey offspring genetically mutated were subjected to painful injections and, in the case of the males, the invasive probe of electroejaculation (apparently at times without anaesthesia). The surrogates into whom the genetically modified embryos were implanted often experienced "difficult births" (Chen et al., 2015, p. 3765), and both mothers and baby monkeys experienced maternal and infant deprivation, which is intensely stressful in mammalian species and known to permanently damage the offspring psychologically.

The fate of the nine monkeys born with the induced muscle cell mutation (which is thought to be the cause of Duchenne Muscular Dystrophy in humans) will likely be more dire than psychological distress. For, if the mutation proves to have the same effect in monkeys as in humans, their lack of a dystrophin gene—"knocked out" by the CRISPR technique of gene-editing—means that their muscles will gradually atrophy and they will die young. The dystrophin gene codes for the protein that maintains muscle cell membranes. In humans Duchenne Muscular Dystrophy occurs only in males: however, in the case with monkeys

Figure 9.1 Baby monkeys with genetically induced muscular dystrophy at 10 weeks. (Reprinted from Chen et al., 2015, p. 3768, with permission from Oxford University Press.)

the induced condition appeared in six females as well (Chen et al., 2015, p. 3769). As the experiment started with 179 gene-edited embryos and only nine usable "models" resulted, there seems to have been considerable "collateral damage" (i.e. dead monkeys or embryos) along the way.

The article describing the experiment includes a disturbing photo of two of the resulting mutated baby monkeys at the age of about 10 weeks. One look at the photo (reproduced in Figure 9.1) gives one a clear idea of how the monkeys feel about their situation. They are frightened, anxious, and sad, clinging to one another for support. Few people could look at this photo without feeling a surge of sympathy for these pitiful creatures. An ethic of care holds that such feelings of compassion should not be ignored, dismissed, or overridden by claims of a higher legitimizing human purpose. Rather, they should serve as ethical guides, alerting us to the inherent immorality and wrongness of this experiment and others of its kind.

REFERENCES

Balcombe, J. 2016. *What a Fish Knows: The Inner Lives of Our Underwater Cousins*. New York: Scientific American/ Farrar, Straus & Giroux.

Cataldi, S. L. 2002. Animals and the Concept of Dignity: Critical Reflections on a Circus Performance. *Ethics and the Environment* 7, 2:104–26.

Chen, Y., Zheng, Y., Kang, Y., et al. 2015. Functional Disruption of the Dystrophin Gene in Rhesus Monkey Using CRISPR/Cas9. *Human Molecular Genetics* 24, 3:3764–3774.

Donovan, J. 2006. Feminism and the Treatment of Animals: From Care to Dialogue. *Signs* 31, 2:305–29.

Donovan, J. 2017. Interspecies Dialogue and Animal Ethics: The Feminist Care Perspective, in *The Oxford Handbook of Animal Studies*, ed. L. Kalof, 208–24. New York: Oxford University Press.

Doudna, J., and S. E. Sternberg. 2017. *A Crack in Creation: Gene Editing and the Unthinkable Power to Control Evolution*. Boston: Houghton Mifflin.

Midgley, M. 2000. Biotechnology and Monstrosity: Why we should Pay Attention to the 'Yuk Factor'. *Hastings Center Report* 30, 5:7–15.

Montgomery, S. 2019. Animal Care. *New York Times Book Review*. 3 March: 1, 16.

Ruddick, S. 1980. Maternal Thinking. *Feminist Studies* 6, 2:342–67.

Schultz-Bergen, M. 2018. Is CRISPR an Ethical Game Changer? *Journal of Agricultural and Environmental Ethics* 31, 2:219–38.

von Uexküll, J. 2010. *A Foray into the Worlds of Animals and Humans*. Minneapolis: University of Minnesota Press.

Weisberg, Z. 2015. Biotechnology as End Game: Ontological and Ethical Collapse in the "Biotech Century". *Nanoethics* 9:39–54.

10

The Privileged Ones[1]

LISA KEMMERER
Professor Emeritus and Founder of Tapestry

Let us imagine a Land of Diverse Communities with the following traits. One group of beings dominates—the Privileged[2] Ones. All non-Privileged individuals and communities are vulnerable in comparison with the Privileged Ones, who have the power to decide the fate of those who are not among the privileged. In the Land of Diverse Communities, the various groups do not share a language and therefore cannot communicate regarding complex matters. However, they are not so different that they cannot understand non-verbal cues—the obvious being expressions of fear or pain: a struggle to escape, crying out, or eyes of terror, as well as such positive expressions as pure joy and playfulness.

Let us further imagine that the Privileged Ones come to believe that they might benefit by exploiting vulnerable individuals from other communities in research, using them as research subjects. They note that sometimes other populations (including vulnerable populations) might benefit from this research, but the end-game is hoped-for benefits for the Privileged Ones. The question then arises among the Privileged Ones as to which vulnerable individuals they might legitimately exploit for their studies, and how they might justify such exploitation. This leads to the related question of who they may *not* exploit, and again, why they may not exploit those particular individuals.

Privileged researchers note that they, the Privileged Ones, are uniquely important—superior—and in any case stand to benefit so greatly that it is essential to exploit the vulnerable. They also note that these other communities are different—not as unique or important, inferior (usually less intelligent or

DOI: 10.1201/9780429428845-10

less cultured). In addition, they remind that vulnerable communities and vulnerable individuals may benefit from the research conducted by the Privileged Ones.

Objectors note that neither they nor individuals of these other communities would choose to be exploited in this manner and that harms to vulnerable subjects exceed hoped-for gains—especially given that the end result may provide not even one hoped-for gain—even for the Privileged Ones. Objectors also argue that the Privileged Ones have core moral rules that they must not break, such as a broad injunction not to exploit the vulnerable for hoped-for/imagined personal gains. Some add that they ought to act from a point of caring, compassion, and connection, and that such exploitation is instead cruel, selfish, and "othering." Finally, objectors note that the best medical models for the Privileged are, obviously, the closest models, which on all counts will be the Privileged Ones themselves.

Alarmed by the idea *they* might be exploited as research subjects, we can readily assume that moral experts among the Privileged might make haste to develop a document on the matter: Let us call this document *Research Protection* (RP) given that it is a document designed to protect "subjects and participants in clinical trials or research studies" (Sims 2010, p. 173) and also to make clear who may (and may not) be exploited as research subjects, complete with protections guaranteed to research subjects. Towards this end, *Research Protection* was designed to identify "basic ethical principles and guidelines" regarding ethical issues that arise from the exploitation of research subjects, with special attention to guidelines protecting any who might be comparatively vulnerable, which would include marginalized Privileged Ones (Summary Statement of The Belmont Report, n.d.).

Let us further suppose that *Research Protection* centres on three principles: respect for individuals, beneficence, and justice. The document that they have created states that the Principle of Respect requires individuals to "be treated as autonomous agents" (*RP* 4). Those "with diminished autonomy," according to *RP*, are "entitled to protection," lest they be exploited (*RP* 4). *Research Protection* also identifies two ethical commitments central to respect for individuals: First, that individuals should be treated as autonomous agents and second, that individuals with diminished autonomy be granted additional protections (*RP* 3).

Also suppose that *RP* states that respect, beneficence, and justice require research subjects to participate voluntarily with volition rooted in informed consent. *Research Protection* goes on to indicate that "the importance of informed consent is unquestioned" (*RP* 7), clarifying that informed consent requires researchers to provide prospective subjects with comprehensive information about proposed research, and that prospective subjects demonstrate an understanding and provide consent via a written or oral test. *Research Protection* acknowledges that some individuals require "extensive protection, even to the point of excluding them from activities which may harm them," while "other individuals require little protection beyond making sure they undertake activities freely and with awareness of possible adverse consequence"—that is, with informed consent (*RP* 4). To be clear, *RP* requires greater protection for "those with diminished autonomy"

(*RP* 4). In these ways, *RP* assures that vulnerable individuals will not and cannot be exploited as research subjects.

Now let us suppose that, based on the Principle of Beneficence, *RP* states that researchers have an *obligation* not to harm research subjects (*RP* 5), including "psychological or physical pain or injury" (*RP* 8) as well as any possible negative effects for their families and/or "society at large (or special groups of subjects in society)" (*RP* 8). In assessing the sufferings of subjects, *RP requires* quantitative and explicit information, not "vague categories" such as "small or slight risk" (*RP* 9). *Research Protection* clearly states that any expected suffering for subjects must "carry special weight" in any moral deliberation (*RP* 8).

While "brutal or inhumane treatment… is never morally justified" (*RP* 9), let us imagine a case where research "involves significant risk of serious impairment," but is found to be in the best interests *of the subject*. In such cases, informed consent and voluntary participation are particularly important, but if such a risky procedure is deemed critical for a vulnerable subject, the privileged assert that *RP* grants subjecting vulnerable subjects to such research—despite the impossibility of informed consent and despite the relative certainty of resultant harms, but *only* once "the nature and degree of risk, the condition of the particular population involved, and the nature and level of the anticipated benefits" *to the research subject* have been *demonstrated* (*RP* 9). Let us suppose that the document is yet more explicit: Such procedures without consent are *only* permissible in cases where suffering borne by the research subject is for the sake of the subject's own benefit and such treatment is only tolerable when research is understood to be essential for the subject's well-being. Let us also imagine that *Research Protection* also notes that, in such cases, a third party may act on the vulnerable subject's behalf, and that the third party is entitled to "observe the research as it proceeds" with the understanding that they are free "to withdraw the subject from the research, if such action appears in the subject's best interest" (*RP* 7).

As we can see, *Research Protection* provides very strong protections to the vulnerable who might be exploited on behalf of those more privileged and powerful. But let us additionally suppose that *RP* asserts that the "selection of research subjects" must be "scrutinized in order to determine" whether or not individuals have been "systematically selected simply because of their easy availability, their compromised position, or their manipulability, rather than for reasons directly related to the problem being studied" (*RP* 6). And let us suppose that this document indicates that it is an *injustice* for research to be conducted on individuals "from groups unlikely to be among the beneficiaries of subsequent applications of the research" (*RP* 6). In other words, the Privileged may not exploit those less privileged or less empowered simply because it is easy and convenient to do so—they must seek research subjects *only* among those who will benefit from the proposed research, *and* they may only accept research subjects who are similar in that they are "directly related to the problem being studied" (*RP* 6). Hoped-for benefits might be taken into consideration, but only "so long as the subjects' rights have been protected" (*RP* 8). In other words, if the privileged are looking for research subjects to use in studies that are expected to largely if not wholly help the privileged, then test subjects must be taken from among the privileged.

In summary, *Research Protections* of the Privileged Ones states that they may not:

- exploit research subjects outside of the community expected to benefit most from proposed research;
- exploit those who cannot give consent—except when the operation is done to benefit the "test subject";
- exploit a vulnerable subject except in the presence of a third party who understands the subject and represents the subject's interests;
- harm subjects, particularly with egregious pain and suffering, unless such harm is demonstrated to be in the subject's best interest;
- exploit subjects based on easy availability, compromised position, or manipulability.

Stated simply, the Privileged Ones may only conduct research on subjects with informed consent from the subject and when research does not harm the subject—unless done for the benefit of the subject—and they must use subjects that are the closest match to those that the proposed research is designed to benefit.

Applied with integrity, *Research Protections* holds very clear indications as to who may and may not be used for research. Given that there is no shared language in the Land of Diverse Communities, it is necessarily impossible for those who are not among the Privileged Ones to provide informed consent. Those communications that are possible do not rise to the level of conveying complex concepts such as "consent." This limits research subjects to those who share their language—the Privileged Ones themselves. Moreover, those intended to be benefited and the closest models (in any definition of the term) are (obviously) the Privileged Ones themselves. According to *Research Protections*, these clear indications and outcomes can only be breached if research is done specifically to benefit the research subject themself.

The Privileged Ones are to be commended for developing such clear and strong guidelines, rooted in respect, beneficence, and justice, to protect the vulnerable from exploitation for research against their will and/or on behalf of others. According to this exemplary *Research Protections* document, may they exploit other communities as research subjects for their ends? Absolutely not.

As you read about the Privileged Ones and *Research Protections*, I trust that you imagined that this scenario was designed to provide answers regarding the suitability of specific research on specific anymals[3] (animals that are not of my species—I am human) such as mice, fish, and macaques. If so, you have imagined incorrectly. The Privileged Ones and their *RP* document are designed to determine whether or not *you* might be used as a research subject.

All information cited above as "*RP*" is from *The Belmont Report: Ethical Principles and Guidelines for the Protection of Human Subjects of Research* (Belmont Report, 1979). This document, published in the United States in 1979, outlines who is eligible as a research subject, and protects vulnerable subjects, exposing *you* as among the vulnerable. If you are a traditional-aged student, for example, you have been vulnerable almost all of your life, incapable of understanding the complex implications of research protocols, incapable of defending yourself against empowered authority figures, and also likely incapable of holding up against bribes or forced servitude. Perhaps one of your grandparents is also vulnerable, or maybe you have siblings or friends-of-the-family who are vulnerable due to medical conditions, or a mental state, whether temporary or permanent. Frankly, even as adults we are vulnerable inasmuch as we are dependent on researchers to explain information in common terms, and fully, and to conduct research in ways that protect the vulnerable, which includes not only research subjects but also consumers. *The Belmont Report* was designed to protect each of us against exploitation by the biomedical industry, government-funded NIH research interests, those who are both desperate and comparatively very wealthy, and any scientists determined to make a name for themselves without regard for ethics, compassion, or respect for life. The above scenario is not about the possibility of protecting mice, fish, or macaques—it is about protecting you.

Do you doubt your vulnerability? The Privileged have certainly exploited "others" as research subjects—thousands of "others"—even in state-funded yet morally abhorrent experiments. For example, at Auschwitz, Josef Mengele exploited twins in the hope of proving the racial superiority of Aryans, while other prisoners were exploited to learn about infectious diseases, the effects of both chemical warfare and extreme temperatures on the human body, and sterilization; Japan's Imperial Army exploited Chinese civilians, causing frostbite to study possible treatments; and male African Americans in the United States were exploited in syphilis studies, allowing them to slowly die long after a cure had been discovered. Private individuals have also exploited the vulnerable as research subjects: James Marion Sims, for example, developed surgical methods for women's reproductive health by experimenting on enslaved, disempowered women of colour and their children—repeatedly, without training, and without anaesthesia—all on behalf of their "owners" (Pappas and McKelvie, 2022, p. 6).

In such cases, the Privileged provide lip-service referencing some morally relevant distinction, however flimsy and bogus, by which they "othered" and exploited their "research subjects." For example, they might have noted that their research subjects did not own property, were from a marginalized community (or class, race, religion, etc.), or that they were deemed to have a lower IQ (or other physical "imperfections"), or perhaps that the test subjects were insensitive to suffering. In each case, the Privileged argued (believed?) that they were in some way (or many ways) special—that they had more of whatever-was-valued by the Privileged, which is to say, they noted whatever might be found to distinguish the Privileged from others. And of course, this was easy to do because the vulnerable *were* different inasmuch as *all* individuals and communities are different in one way or another. Moreover, communication via language was limited/broken between exploiter and exploited. In any event, such exploitation generally passed largely unchallenged because the Privileged Ones held the power, believed that they and their kind stood to gain from such exploitation, believed that they/their lives and welfare were uniquely important—particularly in relation to the lives/welfare of the vulnerable/exploited—and that they and their kind were therefore *entitled* to exploit others for such hoped-for gains.

What about you? Those who are extremely rich and powerful make decisions for/about you every day. They *are* quite capable of exploiting you—you are vulnerable. But does this entitle them to use you for such purposes? Why should those in power be entitled to exploit you for *their* hoped-for gains?

Fortunately, *The Belmont Report*—an actual document—forbids the exploitation of the vulnerable as subjects for research designed to benefit others—even more so when the vulnerable are of a different community, class, race, religion, or IQ (i.e. regardless of intelligence or even consciousness), and if they are viewed as having physical "imperfections," including insensitivity. But as you read, you did not know who the Privileged were, and so you did not know who the vulnerable were. Now you know that *you* are the vulnerable, that the Privileged are perfectly capable of exploiting *you* for research, and that *The Belmont Report* is designed to protect *you* from *their* exploitation.

The Belmont Report was specifically designed to protect *human* beings and has nothing to do with mice, fish, or macaques. Nonetheless, a rat, cat, pigeon, and macaque all fit the category of "vulnerable" and not one of these individuals is capable of providing informed consent as described in *The Belmont Report.* Why should some vulnerable individuals be protected and not others? The time-honoured and core philosophical principles of consistency and impartiality *require* treating similar instances similarly. In light of consistency and impartiality, anyone who fits the description of a vulnerable individual ought to be protected according to the dictates of *The Belmont Report.* Additionally, anyone who does not stand to benefit nearly as much as others must not be forced into the laboratory, whether a child or a chinchilla.

You might reply, "wouldn't *you* kill a mouse to save *your* beloved?" (As if killing a mouse for research could ever guarantee saving a life, let alone a particular life.) This question only serves to highlight the importance of *The Belmont Report*—the importance of protecting the vulnerable. Yes. Fine. I would kill a mouse to save my beloveds—and I would also kill you, and your entire community, and maybe even everyone in every other community, to save my beloveds (including four-leggeds by the way).

If you ask such a question, it is likely that *you* are Privileged: Hearing or reading that you might benefit by exploiting the vulnerable, feeling safe from exploitation yourself, and believing that you *are* more important than the exploited—you stand in favour of exploiting those you view as "other," as "lesser." Those who ask such a question are not qualitatively different from those who supported the research of James Marion Sims and Josef Mengele, those hoping that they might benefit from this cruel and unjust exploitation of others. But of course, that is why members of the medical community came together to create *The Belmont Report*, and that is why other countries have created similar documents—to protect the vulnerable from exploitation on behalf of the selfish interests of the comparatively empowered.

How fortunate that we have a document that clearly describes who may be a research subject and why, as well as why certain others may not be research subjects, and what is owed these vulnerable individuals. More specifically, how fortunate that *The Belmont Report* protects the vulnerable from exploitation as research subjects through the following steps (Table 10.1).

Table 10.1 Overview of *The Belmont Report* protections for vulnerable subjects

Who might be exploited as research subjects?	• Those who provide informed consent via a written or oral document, and primary beneficiaries of proposed research.
Why are certain individuals exploitable?	• Because they are able to provide informed consent and because they stand to benefit directly from the proposed research.
Who may not be exploited as research subjects?	• Those who are unable to provide informed consent and those who will not benefit directly from proposed research.
What are the exceptions?	• Where experiments will not cause egregious suffering and where procedures are conducted on behalf of the research subject.
What is owed all individuals?	• Respect, beneficence, and justice.
What is owed specifically to vulnerable research subjects?	• Research must be done for the benefit of the research subject, and a third party may be present to represent the interests of the vulnerable research subject.

You might object—*The Belmont Report* was only intended to protect vulnerable human beings! True. But those who experimented on "other" human beings in the past (Mengele and Sims, for example) would have posed the same argument, that those they exploited were, at the time and place, outside legal and moral protections for research subjects. History has shown that expedience should never be used to exempt individuals from research protections enjoyed by the privileged.

Moreover, *The Belmont Report* exempts individuals who are incapable of providing informed consent, whether for lack of language or due to an inability to comprehend complex concepts such as "consent." This is obviously the case for anymals, which exempts them from use as research subjects. In any event, *The Belmont Report* requires that research subjects be taken from among those primarily intended to benefit from research conducted, and from among those who stand as the closest models. This also indicates human beings—not fishes and birds and amphibians. Anymals cannot provide consent and they qualify neither as "closest models" nor as primary beneficiaries and so, according to *The Belmont Report*, if anymals are *ever* used in research, it must be for *their* benefit, and the anticipated harms must be unavoidable and neither cruel nor egregious; in such cases, a third party (perhaps a representative from the Physicians Committee for Responsible Medicine or an activist from People for the Ethical Treatment of Animals), must be present during any research to represent the anymal's interests. In light of *The Belmont Report*—according to our own moral assessment and determinations and in light of the time-honoured moral principles of consistency and impartiality—we may not use mice, fish, macaques or any other anymals for our research purposes. According to clear descriptions of the vulnerable in The Belmont Report, anymals qualify as "vulnerable" and are thereby exempt from use as research subjects.

But perhaps you, the reader, are particularly interested in GM research and are wondering: should vulnerable individuals be exempt from protections for research subjects where GM experiments are concerned? Recall that you are among the vulnerable—would you like to be exempt from conscription into vivisection and shock-treatment research but not GM research? Methods of research do not matter—the vulnerable, as described, are protected from being used as research subjects. The question only reveals the importance of *The Belmont Report*.

No methods (including GM), no particular hoped-for gains, no "improved" conditions, and no communities that fit the description of "vulnerable" (including anymals) may be excluded from protections provided in *The Belmont Report* without breaching our own moral guidelines, without putting all who are vulnerable at risk. Of course, this includes GM research that exploits anymals. *The Belmont Report*, when applied consistently and impartially, protects all who are vulnerable– you, me, and all of the anymals (Figure 10.1).

Figure 10.1 It is only because you are privileged that *you* are not in this cage. (Photo of a macaque in a breeding facility, courtesy of We Animals Media.)

NOTES

1 Special thanks to Hope Ferdowsian, who invited me to offer a presentation on anymal research and *The Belmont Report* for PCRM in the spring of 2009, where I first presented this information, though not in narrative form.

2 The concept of "privilege" is important in critical race theory and is therefore indebted to the thought and works of Black scholars and activists. The term is generally used to refer to unfair societal advantages that white people have, or "white-skin privilege," and reaches back to the Civil Rights era.

3 "Anymal" (a contraction of "any" and "animal," pronounced like "any" and "mul"), refers to all individuals who are of a species other than that of the speaker/author. This means that when human beings use the term, they indicate individuals from every species except Homo sapiens. If a chimpanzee signs "anymal," or a parrot speaks the word, individuals of every species (including human beings) are indicated except chimpanzees and parrots, respectively. Using the term "anymal" avoids the use of

- "animal" as if human beings were not animals;
- dualistic and alienating terms such as "non" and "other"; and
- cumbersome terms like "non-human animals" and "other-than-human animals."

See Kemmerer, Lisa. "Verbal Activism: 'Anymals'." *Society and Animals* 14.1 (May 2006): 9–14. http://lisakemmerer.com/Articles/anymal%20article%20Jan%202016.pdf

REFERENCES

Pappas, S and McKelvie, C. 2022. 9 Evil Medical Experiments. https://www.livescience.com/13002-7-evil-experiments.html (Accessed March 24, 2022).

Sims, J. 2010. A Brief Review of the Belmont Report. *Dimensions of Critical Care Nursing*. 29, 4:173–174.

Summary Statement of the Belmont Report. n.d. Ethical Principles and Guidelines for the Protection of Human Subjects of Research. HHS.gov. Office of Human Research Protections. https://www.hhs.gov/ohrp/regulations-and-policy/belmont-report/index.html (Accessed March 15, 2022).

The Belmont Report. 1979. Office of the Secretary: Ethical Principles and Guidelines for the Protection of Human Subjects of Research: The National Commission for the Protection of Human Subjects of Biomedical and Behavioral Research. U.S. Government. https://www.hhs.gov/ohrp/sites/default/files/the-belmont-report-508c_FINAL.pdf (Accessed March 15, 2022).

11

The ethical assessment process

HELENA RÖCKLINSBERG
Swedish University of Agricultural Sciences

DORTE BRATBO SØRENSEN
University of Copenhagen

ANNA KORNUM
Independent Researcher

MICKEY GJERRIS
University of Copenhagen

DOI: 10.1201/9780429428845-11

11.1 INTRODUCTION: ETHICAL REVIEW AND GM

The legal provisions and formal structure of the ethical assessment of animal-based research, in some contexts called ethical review or project evaluation, differs between countries and continents. The responsible bodies for the review, hereafter the Animal Ethics Committees (AECs), are involved to different degrees in approving specific research projects from both public and private sector frameworks (Guillén, 2014). Whereas, e.g., Australia and Canada have no national legislation, as do the EU states, they have comparable well-established institutional ethical review boards since many decades. The equivalent boards in the USA (IACUC, Institutional Animal Care and Use Committee) are regulated by the US national Animal Welfare Act, enforced by the United States Department of Agriculture. This act however excludes birds, rats (*Rattus*) and mice (*Mus*) as well as livestock used for agricultural housing and feeding studies (see Section 3.7.3, and criticism in, e.g., Contreras and Rollin, 2021), and hence limits both scope and responsibility of the ethical review boards' assignment compared to legislation on animal based research in other comparable countries.

In the UK the national authority for project evaluation is the NC3R under the Home Office, but a large responsibility also lies on the additional institutional boards, Animal Welfare and Ethical Review Body (AWERB). In Latin America and Asia one can see large variations in level of regulation. For example, India, South Korea, Taiwan, Malaysia and Taiwan in Asia; and Brazil, Mexico and Uruguay in South America have regulatory systems requiring an oversight of projects in an ethical review board. There are also countries without legal forces, but committees established on voluntary basis, such as Japan, Argentina, Chile, Columbia and Peru. The structure of the boards in Asia and Latin America are often similar to the US boards, but in countries without binding legal framework the boards are challenged to establish enough authority to influence the research along the lines of the 3Rs. (For further information about different approval systems globally, see Guillén, 2014, 2017.)

In this chapter, the main elements of the ethical evaluation process will be highlighted through examples from the European and North American regulation. We have hence no ambition to give a comprehensive global view, but rather to elaborate on factors or dimensions in the approval process that are ethically relevant independently of a specific local process or evaluation system. Simultaneously, through the chosen examples we hope to inspire researchers and AEC members to reflect on both preconditions and ethical principles of an ethical review of animal research as well as on the challenges to implement 3R and a harm-benefit analysis in practice and to learn from systems outside their own.

11.1.1 Different structures of Animal Ethics Committees

In the European Union Directive 2010/63/EU has been the legal basis since 2010, with required implementation before 2013 in each member state. According to the Directive a so-called competent authority shall ensure no projects are approved without ethical assessment, which in practice is often, but not always, delegated to an institutional Animal Ethics Committee which in turn recommends a decision to the competent authority. In contrast to comparison with the US, the number of species defined as research animals is higher since it is based on the criterion of sentience. Chapter IV Article 20 of the Directive 2010/63 states that research projects involving vertebrate animals, foetuses of mammals at two-thirds of gestation and cephalopods require approval. Further, the AEC (or, in some countries, directly, the competent authority) shall perform an ethical evaluation, balancing the societal, educational and scientific benefit of the project with the harm inflicted on the animals. Member states are also obligated to ensure that animal welfare concerns are addressed and that projects comply with the 3Rs (Russell and Burch, 1959).

The federal legislation in the US ad Canada is less detailed than the European Directive. Instead, each state and government may formulate more detailed regulations not only regarding animal welfare but also with respect to the ethical approval system. One example is that although the US Animal Welfare Act does not apply to, e.g., rats and mice (see Section 3.7.3), these species (and many others) are included in the voluntary accreditation system called The American Association for the Accreditation of Laboratory Animal Care (AAALAC). Further, it is worth noticing that the ethical approval system in the US and Canada is regulated on state, not federal, level opening for larger differences between individual states than between EU member states. The Animal Ethics Committees in the US, Institutional Animal Care and Use Committee (IACUC), and the Canadian Animal Care Committees (ACC) have similar structure; they are appointed and responsible at an institutional level, including members from within, such as researchers, veterinarians, animal technicians, non-scientists and students, as well as persons representing local community interests (Schuppli and Ormandy, 2017). In Canada, the ACC's work is guided and overseen by a national organization: Canadian Council on Animal Care (CCAC) stating their own task to "act in the interests of Canadians to 'advance animal ethics and care in science'" (idem, p. 110). Regarding the task, the focus and tool in both US and Canada is the harm-benefit analysis (see the next section) and application of the 3Rs.

11.1.2 Requirements and challenges regarding the ethical review

In practice, however, it has proven difficult for both researchers and AEC members to apply the 3R approach (see, e.g., Schuppli and Fraser, 2005). In a recent Swedish study, it has been shown that the reasons or motivations presented regarding the need to use animals (i.e., not replace them) are not always presented in the application, and that the meaning of refinement and reduction seems to be confused. Further, reduction of numbers is mentioned but often it is not described how this has been considered, and brief or lacking descriptions of how methods are refined. Taken together this frequently make it difficult to follow the actions taken for the different groups of animals used in the actual research protocol. Further, t(his) lack of clarity risks leading to insufficient information to the AECs, which, in turn, not always requires amendments in this regard (Jörgensen et al., Forthcoming 2023).

Both the EU Directive the US and the Canadian regulations request a harm-benefit analysis (HBA) to be performed, balancing the harm done to the animals of the prospective study with the expected benefits for humans, animals or the environment. Part of the review consists of an evaluation of whether the 3Rs have been applied sufficiently since this is fundamental for evaluating the level of potential remaining harm. This task is challenging, not only because most projects consist of several steps with complex procedures involving different groups of animals, relying on assumptions of the level of harm, but also because the future benefit can only be estimated. Although both sides in the balance are hence estimates, the harm typically occurs immediately and in individuals of another species, which makes the welfare assessment – crucial for the evaluation of harm – difficult, whereas the benefit, equally paradigmatically, occurs in the future and for humans. This imbalance in the ethical balancing of harm and benefits is sometimes described as 'a comparison between apples and oranges' (Eggel and Grimm, 2018; Jörgensen et al., 2021).

Further, there is a potential conflict between academic freedom and animal protection, sometime perceived as competing principles (Eggel and Grimm, 2018). On the one hand, researchers wish to conduct research on animals to gain new knowledge, struggling to demonstrate potential tangible benefits of their research, and on the other hand this causes (potential) harm to the animals. Understanding whether the expected future benefits trump the more current harm is thus more than a result of pure calculation of single elements of harm and benefit. Rather it requires a wider ethical elaboration of balancing values – that of fundamental research, of freedom of research, of potential rights of individual animals as well as of animal welfare to name the obvious ones. In addition, wider societal values are at stake, not least when the research involves genetic modification or editing. Values such as trust in medical research continuously contributing to improved medical care, in politicians' priorities of resources, in relevant legislation and its implementation, as well as in the AECs performing an unbiased and skilful evaluation are at the core of the ethical justification at a societal level.

The AECs thus have a complex task of both understanding scientific details in the actual research project and evaluating it in relation to legislation and current societal values and views. In this regard, the EU Directive refers to the importance of the public view. In Recital 12, it is stated that the public considers use of animals in research an ethical issue, and that the intrinsic value and the sentience of the animal shall be respected. Although this is not a legal paragraph, but rather highlights the intention of the legislator, it creates an essential background both for the ethical statements underpinning the claim for replacement whenever scientifically possible, and the expected care of the animals expressed in regulating details of the EU Directive regarding, e.g., handling, housing and humane endpoints.

DIRECTIVE 2010/63/EU, RECITAL 12:

Animals have an intrinsic value which must be respected. There are also the ethical concerns of the general public as regards the use of animals in procedures. Therefore, animals should always be

> treated as sentient creatures and their use in procedures should be restricted to areas which may ultimately benefit human or animal health, or the environment. The use of animals for scientific or educational purposes should therefore only be considered where a non-animal alternative is unavailable. Use of animals for scientific procedures in other areas under the competence of the Union should be prohibited.

Ideally this care for the animals as sentient beings is ensured by the approval process, mainly by the tools of applying the 3Rs and performing the HBA (Recital 39, Article 38). Independently of how or to what degree such elements of a 'public view' requirements are set, society's view remains a core dimension of building societal trust in research, not least in issues related to genetic modification and editing. However, since there is no single 'societal view' on genetic modification of living organisms within EU or a single country, the situation for the AECs is complex. To understand these challenges, the task and role of the AEC's as well as that of the laypersons involved in the project evaluation need to be reflected upon as a web of interrelated issues.

As stated in the beginning, we will here describe the ethical project evaluation in general, elaborate on core challenges such as the HBA and ensuring refinement is applied, i.e., components contributing to striving for an acceptable level of animal welfare, and reflect upon how this task of the AEC is affected using genetic manipulation and editing of research animals. By doing so, we will discuss how AECs may incorporate science, ethics and public opinions into the assessment of project applications.

11.2 THE REVIEW PROCESS: THE HARM-BENEFIT ANALYSIS (HBA)

There are two main challenges related to the ethical review process: (1) the evaluation process as such with the difficulties related to applying 3R and performing a proper HBA, and (2) the transparency of the process and ensuring it meets legal standards. With regard to the first challenge, several international studies show that the AECs seldom, if at all, spend much time on the balancing of harms and benefits, i.e., on the actual ethical elaboration (see e.g. Jörgensen et al., 2021 for a summary of criticism). The reasons are manifold, such as time constraints and disparate views among committee members on the role of the committee (Ideland, 2009), limited understanding of how to apply the 3R (Schuppli, 2004) and insufficient ethical justification (Galgut, 2015). Further, and related, there is frequently a bias for a natural science coloured understanding of ethics limiting the scope of what qualifies as relevant for discussion (Tjärnström et al., 2018). The natural science paradigm of what is considered relevant to discuss, i.e., facts, is often combined with a hierarchical discussion climate structured by the same criteria – representatives of research and natural science are perceived to have the most significant voice also regarding non-scientific and ethical issues. This leads to a limitation of who is listened to and which considerations to prioritize already in the first mapping of what to highlight in the HBA. Hence, the non-scientists have a more difficult position when it comes to influencing the ethical evaluation than that of a researcher (idem; Ideland, 2009, Röcklinsberg, 2015, and McGlacken and Hobson-West, 2022). The most discussed and overarching challenge however concerns the HBA as such. One issue is that views differ between AEC members on how to understand harm and benefit respectively, and it is hence challenging to agree upon a common definition. Another issue is that balancing and evaluating harms vs benefits in a 'proper' HBA is no easy task. These challenges will be elaborated on in the following, whereas the second main challenge, transparency and legality, will be discussed briefly in Section 11.3.

11.2.1 Models of harm-benefit analysis

Due to the complexity of performing a HBA it has been the object of much concern in the interface between lab animal science and applied ethics for many years (see, e.g., Bateson, 1986; Stafleau et al., 1999; Bout et al., 2014; Grimm et al., 2017; Ringblom, 2017; Laber et al., 2016; Eggel and Grimm, 2018, and Gerritsen, 2022 for a variety of approaches including legislative aspects). The variety of suggested models is too large to map here, and the following presentation will focus on three, in order to shed light on different dimensions and challenges.

11.2.1.1 THE BATESON CUBE AND REVISED VERSIONS

The perhaps most influential model, the so-called Bateson Cube (Bateson, 1986, 2005), was presented already in 1986, serving as the basis for further development and nuancing of ethically relevant considerations. It was developed by Patrick Bateson, himself an animal behaviourist criticized by anti-vivisectionists, as a tool to facilitate dialogue between advocates and critics of animal-based research, and as an attempt to meet public concerns. It was timely launched in the preparation of the new British legislation on animal research, Animals (Scientific Procedures) Act 1986. The cube form expresses the three dimensions of evaluation of all research involving animals, each on a scale from low to high: (1) the degree of animal suffering, (2) the likelihood of benefit (quality of the research) and (3) the importance of research. Only projects fulfilling all three dimensions, ideally at a high level, are ethically justified and possible to approve by an AEC. Hence, the dimensions can be interpreted as requirements, all holding the same level of relevance (Figure 11.1).

A modern version of the Bateson Cube, also intended to facilitate dialogue within the AEC and with researchers, is a matrix model developed by Bout et al. (2014). They insert a distinct difference to Bateson regarding the equality of the requirements: the likelihood of reaching the potential benefit should be a fundamental requirement for all research.

> Our proposed matrix treats the likelihood of achieving the benefits not as part of the weighing process in the harm-benefit analysis but instead as a prerequisite: if the animal use proposal has insufficient

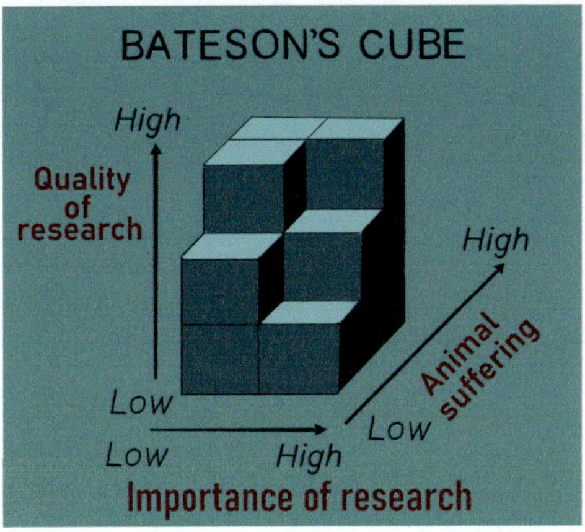

Figure 11.1 The Bateson Cube illustrates the three dimensions he argues should be assessed and approved in order to ethically justify animal-based research: (1) the degree of animal suffering, (2) the likelihood of benefit (quality of the research) and (3) the importance of research. Later research suggests a hierarchy between these dimensions, requesting either the likelihood of the research to reach envisaged benefits to be fulfilled before the other dimensions are evaluated (Bout et al., 2014) or the quality of research (see below (Illustration Anna Kornum)).

> likelihood of achieving the benefit, then the outcome of the ethical review must be negative, rendering a further harm-benefit analysis unnecessary.
>
> *(Bout et al., 2014, p. 413)*

The suggested HBA to follow ought to distinguish between mild, moderate and severe harms to the used animals, and divide benefits into the seven purposes of using animals in science as stated by the Directive; basic research, translational or applied research, product safety, education and training, protection of the environment, preservation of species and forensic inquiries (article 5, 2010/63/EU). Bouts et al. suggest that these purposes together mirror the potential benefits, which may be divided into those of a more scientific and social character respectively.

This matrix model can be seen as a means to highlight the challenges in formulating 'the optimal model' for an ethical review of animal research. Hence, the points made in the following applies to any other similar model. Although it is reasonable to require an assessment of what is a 'sufficient likelihood' of achieving a benefit, it might be a very difficult task. The AEC would need to clarify along what criteria this likelihood of social and scientific benefits would be assessed, as well as for whom and in what time frame. In principle the potential benefits are almost unlimited which calls for an ethical elaboration of when likelihood is achieved.

For example, researchers conducting studies of specific genes or interaction between genes in a mouse model designed for a common human disease such as obesity envisage new insights that are directly corresponding to the study's hypothesis. Such a study is likely to be regarded as beneficial by an AEC. However, given the numerous studies of this kind, and further relevant aspects such as the complexity of the causes of obesity, one may still wonder if it is *sufficiently* likely, or unlikely, to contribute to something valued as beneficial. There are thus still several evaluative steps to be taken by the AEC, and a set of criteria need to be developed to decide whether to classify such an obesity study as likely leading to enough beneficial output or not.

Further, it is relevant to remember that much research on genes and their expressions is performed as basic research, for the sake of understanding biological fundamentals, which per se makes the evaluation of likelihood of benefits difficult. They can simply not be known beforehand. The discovery of the 'genetic scissor' CRISPR/Cas9, albeit not by using animals, is the result of fundamental research and the method of gene editing would perhaps not have been developed if approving the likelihood had been the first needle eye to pass. Whether this is a bad or a good is of course a normative question, but it seems possible to conclude that this requirement of likelihood of achieving proposed benefits can be interpreted as one way to limit fundamental research, not only regarding genetic modification and editing, but without doubt, including it. (How the AEC may evaluate fundamental research within their HBA will be further elaborated below.)

On a more structural note, Bout et al.'s model establishes a hierarchy between the three requirements: likelihood of achieving proposed benefit (quality of research) as a prerequisite, followed by assessing the importance of research (scientific benefits and social benefits, respectively) and harm to animals. This is a promising approach, but it is worth reflecting upon whether stressing another part of the requirement might be the most a more suitable prerequisite. An alternative reformulation of the Bateson's cube to such a two-step process could be to further stress the 'quality of the research' as a necessary condition for all further ethical evaluations. That is, a thorough weighing of harms and benefit need to be performed *only* if the suggested research design and overall quality is guaranteed. The reason is obvious. Results derived from faulty protocols with unknown side effects, or a study design not corresponding to the said hypothesis, or with low or no translational value are simply useless and thereby irrelevant (Ioannidis, 2005 and Garner et al., 2014). Würbel (2017:165) lists some illustrative examples of pitfalls in the scientific planning and conduct.

> For example, studies may use poorly validated animal models or outcome variables; they may be based on samples that are too small or idiosyncratic; they may violate principles of good research practice (for example, randomization, blinded outcome assessment, a priori sample size calculation) or use inappropriate statistics (for example, *p*-hacking); or they may report results selectively or not at all (for example, publication bias).

If the research hampers with any of these pitfalls, animals and resources have been used in vain and can be understood as a sign of low research ethics standards and limited awareness of one's responsibility while conducting research (Röcklinsberg et al., 2017). It is therefore crucial that the AEC first ensures that the scientific method proposed has a high potential to produce valid and repeatable results, i.e., build on good scientific quality, before spending time and efforts in balancing the actual harms and potential importance of research benefits.

Although societal trust in science is a complex issue, it is safe to say it would be at risk if too much research were lacking scientific quality (for an elaboration on social relevance, see Meijboom, 2017). Hence, a responsible approach to research begins long before the actual experiment is executed.

Finally, another argument for raising the relevance of research quality as a core focus preceding the claim of likelihood of achieving expected benefits lies in the possibility to evaluate the fulfilment of the requirement. A (close to) objective assessment of whether a proposed research design is solid is possible via well-established tools such as statistical power analysis, proper use of control groups, choice of relevant animal models, accurate housing and handling, as well as scrutinizing whether the 3Vs (construct validity, internal and external validity) (Würbel, 2017) are implemented. Hence, in relation to HBA the stepwise evaluation process would be that only if there are good reasons to trust that the requirements of the 3Rs are met, the proposed research can be expected to hold a sufficient qualitative level which in turn opens for the next step of the ethical review, i.e., animal welfare assessment and evaluating potential benefits. As clearly highlighted by the 'biomedical crisis', scientific methods can hence be well chosen and studies well performed, or not. It is the task of the AEC members to use its scientific skills to scrutinize the quality of the design to ensure the research meets standards of validity and repeatability, and implicitly animal welfare, in order to fulfil the overarching ethical claim regarding scientific quality (see, e.g., the Directive 2010/63/EU Recital 13, or American Association for the Accreditation of Laboratory Animal Care (AAALAC)). Such an evaluation will show whether an animal-based or animal-free research design is the most suitable.

11.2.1.2 THE AALAS-FELASA MODEL OF HARM-BENEFIT ANALYSIS

In parallel to the reflection on the Bateson Cube above, we will in the following section elaborate on challenges related to a recent HBA model to invite reflections on some challenges any AEC will meet at the very core of their task, independently of which specific process a specific committee has chosen, as described in the scientific literature on the performance of the AECs.

Of special interest for the performance of a proper HBA is the suggestions made by the American Association for Laboratory Animal Science–Federation of European Laboratory Animal Science Associations (AALAS-FELASA) working group (Laber et al., 2016). The authors state the aim is to establish "a systematic process for HBA [which] is expected to ensure that all potential harms and benefits have been comprehensively and carefully considered during the ethical evaluation of the merits of an animal research investigation" (p. 21). It shall further ensure the scientific value of each research step for testing the hypothesis and evaluate each component of the procedure with regard to the harm inflicted.

The working group suggests that harms and benefits are divided into domains, each including several sub-areas. Worth noting is that harm is explicitly defined to be a public concern and understood as a wider concept than pain alone (p. 22), while also being defined along the lines of *The Five Freedoms* (see Chapter 4 on Welfare and 3R in this book). Subdomains of benefits listed by Laber et al. (2016) are social, socioeconomic, scientific, educational, and safety and efficiency benefits. To facilitate a transparent and comprehensive evaluation they suggest a step-by-step procedure following a template along a number of 'modulating factors'. These are, e.g., animal species and number, experimental intensity and duration, genetic modulation and housing in the harm domain. Modulating factors in the benefit domain are, e.g., clarity of objective, translational potential and innovation level. The task of the AEC is to estimate the weight of harm and benefit in each factor, and when done so marking it with a colour (pail to intensive) or numbers, in order to facilitate a visualization of the judgement of the factors. This, it is stated, will lead to a conclusive, transparent and solid balance of harm and benefits.

It can be observed that although the approach is well described and exemplified by a filled-out coloured template, it most probably remains challenging for many AECs. According to several studies on the performance of the AECs the first obstacle to overcome is seeing the value of and need to even establish a routine of respectful ethical discussions between the committee members. This is due to the fact that many AECs suffer from time constraints, a lack of consensus regarding the role of an AEC and hierarchies between scientists and laypersons, leading to the latter group to being hesitant to contribute (see, e.g., Schuppli, 2004; Sengupta, 2003; Ideland, 2009; Varga, 2013; Röcklinsberg, 2015; Tjärnström et al., 2018; Jörgensen et al., 2021). Such factors inevitably have negative implications for the possibility of the AEC reaching agreement on how to estimate harm and benefit for the numerous 'modulating factors' in the HBA model. Given such difficulties in establishing a well-functioning routine of ethical elaboration, it can be nearly impossible to create an open discussion climate, and moreover, the lack of openness implies that fewer relevant aspects – as listed in the AALAS-FELASA model – are considered in the AEC decisions. Further, this leads to a risk of not mirroring public concerns of where to draw the line between necessary and unnecessary animal suffering in relation to relevant and non-relevant research. These challenges are not only severe and long lasting in themselves but also highlight the need to facilitate discussion of *ethical* dimensions, above technical and administrative, regarding animal-based research.

One of the main challenges *within* an ethical elaboration regards estimation of the benefits of new research methods or tools, such as genetic modification or genome editing. Here a final suggestion for the HBA process is worthy of mentioning, since it sheds some new light on what is understood as 'benefit' and leads back to the issue of assessment of the value (benefit) of fundamental research, including developing methods for and applying genetic modification, elaborated on above.

11.2.1.3 FROM AN HBA MODEL TO A HARM-KNOWLEDGE ANALYSIS

Eggel and Grimm (2018) suggest an HBA model where the evaluation components shift from benefit to scientific knowledge and the evaluation content is reduced in terms of which questions shall be covered by the AEC. The latter theme will be included below (Section 11.3), at first reflecting on the suggested component of the HBA. According to them, the concept of harm has been adequately defined whereas that of benefit has not. By implication it is then difficult to use properly in an HBA. Moreover, they emphasize the impossibility of predicting the actual long-term, or ultimate, societal benefits of the knowledge gained in many animal experiments. There is a risk that the current practices of AECs might encourage applicants to simplify their research or overstate the societal benefits of their projects in order to obtain approval. Please note that this argument also pertains when following the matrix by Bout et al. (2014) mentioned above when estimating the likelihood of benefits as a prerequisite for further HBA. Consequently, Eggel and Grimm argue that scientists who are involved in basic research have a difficult time showing the ultimate benefits of their research. This creates a bias, to the disadvantage of fundamental research increasing general biological or medical knowledge. Eggel and Grimm argue these societal benefits of animal-based research can and should be assessed retrospectively and suggest a shift to a harm-*knowledge* analysis: focus on benefit as knowledge would reflect a more realistic notion of how benefits of animal research are achieved, more often than not via fundamental research which seemingly has no societal value. Overall increased knowledge is the ultimate aim for research, and the application of it into further research and eventually medical treatment.

Similar to the challenges of the Bout et al. model of estimating likelihood, it seems relevant to ponder upon whether it is easier to predict knowledge gain than any other (scientific or social) benefit. Further, it seems challenging to evaluate overall social benefits retrospectively and for all kinds of research in a time frame that can feed back relevant information to the AECs and researchers about which study design was most successful *before* further projects are formulated. Feedback on scientific benefits will be easier, but even so, the question arises how that will justify the use of animals harmed in the trials, if discovered afterwards that the benefit was low? Hence, it seems that retrospective assessment may well be valuable, but if taken seriously, it implies a slower process from finalized research to the researchers' next step to formulate and apply for a new AEC assessment, and that the project evaluation shall not be performed until the retrospective assessment of benefits is performed, interpreted and reflected upon among researchers in the field. This most probably would lead to improved study design, increased translation rates and increased implementation of all the 3Rs but increase the time between application and approval by an AEC.

These three suggestions of how to perform a 'proper' HBA (the Bateson Cube, incl. the Bout et al. version of it, the AALAS-FELASA score sheet and the version suggested by Eggel and Grimm) could, when applied, facilitate the challenging work of the committee members. Also, it may lead to insight into the inherent ethical issues in performing research on one sentient being for the sake of another, of evaluating 'moving targets' such as unknown consequences of genetic modification or levels of reduced welfare, of potential negative side effects or confounding factors influencing the result, and of per se unknown details regarding the results, i.e., potential benefits – since it is research – and of often limited time frame of the study. The presentation of the HBA-analysis models, however, also highlights the dire need for a guideline, or a step-by-step process description, to make the complex task of the committees feasible and facilitate reaching well-justified decisions. The discussion about the models is further inevitably linked to the issue of the committee composition, i.e., who shall participate in the ethical evaluation of animal-based research, including the assessment of genetic modification, and to the issue of the role of the public, and how to take 'public concern' into account (Hobson-West, 2009; Meijboom 2017; Lassen and Bøker Lund 2017; McGlacken and Hobson-West, 2022).

11.3 WHO PARTICIPATES IN THE ETHICAL REVIEW ON ANIMAL RESEARCH?

In North America, Latin America, Eurasia and Oceania scientists are assigned a central role in the AECs. Often researchers are in the majority, sometimes with a low number of laypersons, and in addition there is often a lawyer to ensure legislation is correctly implemented (Guillén, 2014, 2017). The EU Directive leaves the question of how to structure the AEC ('the competent authority') open for each member state to decide. The Swedish system is unusual by having equal number of researchers and laypersons in the six regional board (six+six), handling all applications from universities and research facilities in a region. Four to five laypersons are appointed from the political parties in the main municipality of the region and one to two laypersons from an animal welfare organization. The former shall not represent any political standpoint, but rather society in general, whereas the latter are expected to have a special focus on animal welfare. Although this 50-50 structure supposedly ensures a public view is taken into consideration while establishing a broad basis for the ethical elaboration and HBA, it has long proven difficult in practice (Forsman, 1992; Ideland, 2009; Tjärnström et al., 2018). The ideal of high layperson involvement has, however, recently inspired the medical faculty at the Johannes Kepler University of Linz, Austria, to structure its institutional committee in a similar way (Diabl, 2022).

Lay member involvement raises several questions: Is lay involvement for reasons of rationality, moral or democratic legitimacy; what is the reason, in the first place, that the public should be involved in the ethical review; and whom are they supposed to represent? (e.g., Stafleau et al., 1999; Hobson-West, 2009; McGlacken and Hobson-West, 2022). There are several groups, some of which have representatives in the North American systems (non-scientists and community members) and Swedish committees (animals and the public), whereas others such as patient groups and representatives for nature as such are not. This leads to more detailed questions of what is expected from the lay members (which is often in contrast to what they experience (e.g., Tjärnström et al., 2018)) and what it takes to ensure it is possible to exectue.

Eggel and Grimm (2018) argue in favour of a public and political debate and elaboration on the different forms of benefits, suggesting the responsibility to set the limits of what is acceptable, and what not, in a society, lies on politicians and the general public, not primarily on the AEC. This is a tempting approach in terms of potentially increased public involvement and democratization of the ethical deliberation on the aims and methods of animal research. Such a system would, however, probably be challenging to implement. Any public discussion would run the risk of being restricted to build on the limited information presented in the 'non-technical summaries' in order not to reveal internal research secrets, and hence most of the actual research process and conduct will not be included in the general knowledge about the research at stake. Further, as only the general principles behind using animals in research are to be up for discussion, regarding, for example, animal species, housing facilities and genetic engineering, this is nothing new in the public discourse, rather it is ongoing in many countries. For example, the above-mentioned Medical Faculty in Linz aims to establish medical research with as minimal use of animals as possible, to fulfil the Directive of phasing out animal-based research (Diabl, 2022).

Further, strong signs of public engagement in phasing out, and meanwhile improving application of the 3Rs in animal-based research, is the Citizen initiative on improved implantation of the Directive of 2021. Over 1.4 million signatures were collected leading to an EU Parliament decision on September 16, 2021, to increase the speed in phasing out animals in research, and to raise the ambitions regarding animal welfare (EU Parliament P9_TA(2021)0387, n.d.). This statement is most relevant with respect also to the specific ethical issues related to genetic modification and editing and the public discussion, not least in the light of the questions arising regarding the translational value of animal models.

The public gains increased knowledge about alternatives to animals, such as organ on a chip and in silico methods but cannot be expected to know about the pros and cons of these new methods, nor the differences in outcome compared to animal-based methods in relation to specific research aims. Further, researchers and critics evaluate the limits and potentials differently, partly due to their view on what ethical issues are at stake regarding the spectrums of (1) animal welfare vs animal rights/integrity and (2) benefits/risks vs 'naturalness'/animal rights regarding GM/GE and (3) maximizing 'benefit' (however defined) vs limits to the responsibility of the researcher. The intersections of these issues lead to different views on what is ethically acceptable to conduct with the animal and by the researcher (see Chapter 4 for a more elaborate discussion of the complexity of ethical deliberation).

Here, point 2 is of special interest, since it concerns the perception of what is ethically relevant in terms of genetic modification. Röcklinsberg, Gamborg and Gjerris (2014) argued that harm and benefit are not all that matters ethically: broader ethical concerns, such as integrity, should be given more consideration in AEC assessments. According to the authors, in the HBA animal welfare is often not considered equally relevant compared with potential human benefits, and discussion of ethics is often narrow, superficial or even absent within the assessment process. Although animal welfare is often given considerations, the concept of animal welfare is itself embedded in a larger ethical framework (Fraser, 1997), which the AECs must be aware of. Broader ethical concerns such as the integrity of the animal and naturalness might not have any bearing on the subjectively experienced welfare of the animals themselves, but might matter ethically, and related ideas of naturalness are shown to play a role to the public especially in relation to genetic modification (see Chapter 4). If AECs ignore such ethical concerns on harm, in combination with limiting their understanding of welfare to biological and affective dimensions, they risk becoming alienated from public expectations, in particular in cases where genetic manipulation and editing is involved. An ethical assessment process with a multidisciplinary approach, taking the range of ethical aspects serious, is better served to reach a solid statement. Further, a democratic approach involving the opinions of both experts and laymen will more easily ensure public legitimization for the use of animals in experiments.

This position lies close to the view presented by Hawkins and Hobson-West in their elaboration on the role of the British Animal Welfare and Ethics Review Body (AWERBs) (Hawkins and Hobson-West, 2017), based on a description of common challenges establishing public trust, mentioning the ethical elaboration as one specific challenge. All applications in the UK are assessed by a national committee at the Home Office, different in dimension and by hosting a number of persons performing the evaluation as their profession, but similar to the AECs discussed in this chapter so far, being the assigned national authority for project evaluation. Hence, the AWERB shall only suggest a decision, but has many tasks close to those of an animal welfare body (Article 27 of the Directive 2010/63/EU), in terms of advising and informing researchers and technical staff on alternative methods, refinement and reduction, evaluating routines in the light of animal welfare, following-up projects and giving advice regarding rehoming schemes.

The British structure for project evaluation was established while part of EU and a core partner in the elaboration of the Directive, but differs in many respects from the Swedish structure. This constitutes, together with the FELASA group HBA model which was developed by researchers from both North America and EU, good examples of the benefit of openness regarding format of the ethical project evaluation while aiming to fulfil the legally requested content. The challenges in establishing an ethical discussion within which a proper HBA is performed and the 3Rs fully implemented however remain largely the same.

11.4 CONCLUSIONS AND IMPLICATIONS

While t(here) seems to be a global acknowledgement of the inherent ethical dimensions of using animals in research for the sake of human medical or other purposes, there are numerous ways of handling the evaluation of these dimensions, especially with regard to legislation, the evaluation process and whose interests are included. Similar to many issues in applied ethics, there might be no 'true' or *one optimal* answer serving all situations equally well. Rather, there is a plethora of elements to consider: the challenges in meeting contradicting criteria – such as reducing the number of animals while breeding numerous surplus individuals when striving to achieve a certain line of genetically modified mouse – and the increased biological and ethological knowledge about what an animal actually perceives. All this may be influenced by – and hence will have an impact on – the result and calls for a discussion of what is at the core of the ethical evaluation by the AEC. Is it the precise assessment of the harm done to the animal, the potential societal value of a forthcoming medical treatment or of fundamental biological knowledge, or the validation of the scientific quality of a certain experiment? The answers to these questions are dependent on values. The values of individual researchers, committee members and politicians will influence what they argue for, as well as which HBA model they prefer. In order to handle conflicting values clearly stated requirements regarding three issues is crucial: how to achieve a proper ethical evaluation, a well-argued decision based on a transparent evaluation process, and knowledge about different models.

The more complex the research protocol is, for example, by using an unknown strain of genetically modified animals, the more important it is that the AEC holds a high standard in applying the legislation and guidelines set, ensuring strong ethical competence. That is, the individual AEC member's ethical competence building lies in the very core of ensuring each evaluation and decision made is justified, almost independently of which the HBA model is used. Hence education of AEC members' ethical and legal skills and chairpersons' competence as discussion leaders is at the core of an open elaboration mapping of all relevant aspects, grading their relevance and comparing them to each other in order to enable an ethically justified decision deserving public trust.

REFERENCES

Animals (Scientific Procedures) Act. 1986. https://www.legislation.gov.uk/ukpga/1986/14/contents (accessed March 12, 2023).

Bateson, P. 1986. When to Experiment on Animals. *New Scientist* 109:30–32.

Bateson, P. 2005. Ethics and Behavioral Biology. *Advances in the Study of Behavior* 35:211–233.

Bout, H., van Vlissingen, J.M.F. and E. D. Karssing. 2014. Evaluating the Ethical Acceptability of Animal Research. *Laboratory Animals* 45, 11:411–413.

Contreras, E.T. and B.E. Rollin. 2021. The Convenient Disregard for the *Rattus* Species in the Laboratory Environment: Implications for Animal Welfare and Science. *Journal of Animal Ethics* 11, 2:12–30.

Diabl, C. 2022. JKU Linz startet 2023 mit Tierversuchen unter strengen Auflagen. Oberösterreich Nachrichten (news) (Sept 26, 2022). https://www.nachrichten.at/oberoesterreich/tierversuche-an-der-jku-starten-2023;art4,3717572 (accessed February 15, 2023).

Eggel, M. and H. Grimm. 2018. Necessary, but Not Sufficient. The Benefit Concept in the Project Evaluation of Animal Research in the Context of Directive 2010/63/EU. *Animals* 8, 3:34.

EU Parliament P9_TA(2021)0387. n.d. Plans and actions to accelerate a transition to innovation without the use of animals in research, regulatory testing and education. https://www.europarl.europa.eu/doceo/document/TA-9-2021-0387_EN.html (accessed Feb 21, 2023).

Forsman, B. 1992. *Djurförsök: Forskningsetik, Politik, Epistemology: En Vetenskapsteoretisk Kontextualisering*. Göteborg, Sweden: Almqvist & Wiksell International.

Fraser, D., Weary D. M., Pajor, E. A. and B.N. Milligan. 1997. A Scientific Conception of Animal Welfare that Reflects Ethical Concerns, *Animal Welfare* 6: 187–205.

Galgut, E. 2015. Raising the Bar in the Justification of Animal Research. *Journal of Animal Ethics*, 5, 1:5–19.

Garner, J.P., Gaskill, B.N., Weber, E.M., Ahloy-Dallaire, J. and K.R. Pritchett-Korning. 2014. Introducing Therioepistemology: The Study of How Knowledge is gained from Animal Research. *Laboratory Animals* 26, 4:103–113.

Gerritsen, V. 2022. *Güterabwägung in Tierversuchs bewilligungsverfahren. Series on Animal Law and Ethics, Volume 23*. Zürich: Schulthess Publishing Company.

Grimm, H., Eggel, M., Deplazes-Zemp, A. and N. Biller-Andorno. 2017. The Road to Hell Is Paved with Good Intentions: Why Harm–Benefit Analysis and Its Emphasis on Practical Benefit Jeopardizes the Credibility of Research. *Animals* 7, 9:70.

Guillén, J. 2014. *Laboratory Animals: Regulations and Recommendations for Global Collaborative Research*. Amsterdam: Elsevier/Academic Press.

Guillén, J. 2017. Regulation and Review of Animal Experiments in Asia, Latin America an Oceania. In *Animal Ethics in Animal Research*, eds. H. Röcklinsberg, M. Gjerris and I.A.S. Olsson, 119–124. Cambridge: Cambridge University Press.

Hawkins, P., and P. Hobson-West. 2017. *The AWERB as a forum for discussion*. https://view.pagetiger.com/AWERB/AWERB (accessed March 12, 2023).

Hobson-West, P. 2009. The Role of Public Opinion in the Animal Research Debate. *Journal of Medical Ethics* 36:46–49.

Ideland, M. 2009. Different Views on Ethics: How Animal Ethics is Situated in a Committee Culture. *Journal of Medical Ethics* 35:258–261.

Ioannidis, I.P.A. 2005. Why Most Published Research Findings are False. *PLoS Medicine* 2, 8:e124.

Jörgensen, S., Lindsjö, J., Weber, E.M. and H. Röcklinsberg. 2021. Reviewing the Review: A Pilot Study of the Ethical Review Process of Animal Research in Sweden. *Animals* 11, 3:708.

Jörgensen, S., Lundmark-Hedman, F., Lindsjö J., Weber E.M. and H. Röcklinsberg. Forthcoming. Debilitating Dissimilarities Disrupting Delicate Decisions (2.0).

Laber, K., Newcomer, C.E., Decelle, T., Everitt, J.I., Gullien, J. and A. Brønstad. 2016. Recommendations for Addressing Harm-Benefit Analysis and Implementation in Ethical Evaluation—Report from the AALAS-FELASA Working Group on Harm-Benefit Analysis—Part 2. *Laboratory Animals* 50:21–42.

Lassen, J and Bøker Lund, T. 2017 Public Involvement: How and Why? (Section 6.1 and 6.2) In *Animal Ethics in Animal Research*, ed. H. Rocklinsberg, M. Gjerris and I.A.S. Olsson, 139–151. Cambridge, UK: Cambridge University Press.

McGlacken, R. and P. Hobson-West. 2022. Critiquing Imaginaries of 'The Public' in UK Dialogue around Animal Research: Insights from the Mass Observation Project, *Studies in History and Philosophy of Science* 91:280–287.

Meijboom, F.L.B. 2017. Applying Ethical Thinking and Social Relevance. In *Animal Ethics in Animal Research*, ed. H. Röcklinsberg, M. Gjerris and I.A.S. Olsson, 68–90. Cambridge, UK: Cambridge University Press.

Ringblom, J. 2017. Assigning Ethical Weights to Clinical Signs Observed During Toxicity Testing. *ALTEX* 34:148–156.

Röcklinsberg, H. 2015. Lay Persons Involvement and Public Interest. Ethical Assessment in Animal Ethics Committees in Sweden. The Swedish Transition Process of the EU Directive 2010/63/EU With Regard to Harm-Benefit Analysis in Animal Ethics Committees. *ALTEX Proceedings* 4, 1:45–48.

Röcklinsberg, H., Gamborg, C. and M. Gjerris. 2014. A Case for Integrity: Gains from Including More Than Animal Welfare in Animal Ethics Committee Deliberations. *Laboratory Animals* 48, 1:61–71.

Röcklinsberg, H., Gjerris, M. and I.A.S. Olsson. 2017. *Animal Ethics in Animal Research*. Cambridge, UK: Cambridge University Press.

Russell, W.M.S., and Burch, R.L. 1959. *The Principles of Humane Experimental Technique*. London and Methuen, MA: Universities Federation for Animal Welfare (UFAW).

Schuppli, C.A. 2004. The role of the animal ethics committee in achieving humane animal experimentation. Ph.D. Thesis, University of British Columbia: Vancouver, British Columbia, Canada.

Schuppli, C.A. and D. Fraser. 2005. The Interpretation and Application of the Three Rs by Animal Ethics Committee Members. *Alternatives to Laboratory Animals* 33:487–500.

Schuppli, C. and E. Ormandy. 2017. Regulation and Ethics Review in North America. In *Animal Ethics in Animal Research*, eds. H. Röcklinsberg, M. Gjerris and I.A.S. Olsson, 109–119. Cambridge,: Cambridge University Press.

Sengupta, S. 2003. The Roles and Experiences of Non-Affiliated and Non-Scientist Members of Institutional Review Boards. *Academic Medicine* 78:212–218.

Stafleau, F., Tramper, R., Vorstenbosch, J. and J.A. Joles. 1999. The Ethical Acceptability of Animal Experiments: A Proposal for a System to Support Decision-Making. *Laboratory Animals* 33:295–303.

Tjärnström, E., Weber, E., Hultgren, J., and H. Röcklinsberg. 2018. Emotions and Ethical Decision-Making in Animal Ethics Committees. *Animals* 8: 181.

Varga, O. 2013. Critical Analysis of Assessment Studies of the Animal Ethics Review Process. *Animals*. 3:907–922.

Würbel, H. 2017. More than 3Rs; the Importance of Scientific Validity for Harm-Benefit Analysis of Animal Research. *Lab Animal Commentary* 46:164–166.

Appendix

Statutes and regulations in selected countries and geopolitical units affecting animals produced by modern biotechnology

Country/ geopolitical unit	Laws/regulations specifically addressing animals produced by modern biotechnology	Specific issues addressed by laws and regulations
European Union	Laws and regulations specific to production of GE animals; laws and regulations specific to cloned animals; also, laws and regulations exist that are specific to welfare of GE animals and cloned animals	GM ANIMAL PRODUCTION • EU regulates the process of genetic alteration and use to produce food products or ingredients from GM animals (REGULATION (EC) No 1829/2003). • EU grants "authorization" for GMOs as source materials for food production or use with premarket safety assessment by European Food Safety Authority (EFSA). • EFSA guidance specific to assessing the risks to human health posed by GM animals (https://www.efsa.europa.eu/en/efsajournal/pub/2501) December 14, 2011. • EFSA published guidance specific to GM animals and their impact on the environment (https://www.efsa.europa.eu/en/efsajournal/pub/3200) April 18, 2013. • The European Medicines Agency (EMA) issues guidance on use of GM animals to produce pharmaceuticals and for transgenic animal models used to test compounds (*"Use of Transgenic Animals in the Manufacture of Biological Medicinal Products for Human Use" 1995*). • EU Directives 2001/18/EC (deliberate release of GMOs into the environment) and Directive 2009/41/EC apply to production or incorporation of GMOs into medical therapies. • July 25, 2018, Ruling by the Court of Justice of the European Union classified gene-editing in animals as a GMO (https://curia.europa.eu/jcms/upload/docs/application/pdf/2018-07/cp180111en.pdf).

(Continued)

Statutes and regulations in selected countries and geopolitical units affecting animals produced by modern biotechnology

Country/ geopolitical unit	Laws/regulations specifically addressing animals produced by modern biotechnology	Specific issues addressed by laws and regulations
		ANIMAL CLONE PRODUCTION • EU considers animal cloning separately. • Foods from animal clones are considered to be "Novel Foods" and must undergo a premarket assessment. • European Parliament bans the cloning of all farm animals, sale of livestock clones, their offspring, and products derived from them. The ban does not cover cloning for research purposes, nor does it prevent efforts to clone endangered species. • Directive issued relates to the internal EU market only, NOT IMPORTS to the EU. • Member States will now have to address this issue on a country-by-country basis in terms of regulations to develop. GM ANIMAL WELFARE • Directive 2010/63/EU. • *2012 Guidance on the risk assessment of food and feed from genetically modified animals and on animal health and welfare aspects.* EFSA Panels on Genetically Modified Organisms (GMO) and Animal Health and Welfare (AHAW). EFSA Journal 2012. 10(1):2501. • *2013 Guidance on the environmental risk assessment of genetically modified animals.* EFSA Panel on Genetically Modified Organisms (GMO). EFSA Journal 2013. 11(5):3200.

(*Continued*)

Statutes and regulations in selected countries and geopolitical units affecting animals produced by modern biotechnology

Country/ geopolitical unit	Laws/regulations specifically addressing animals produced by modern biotechnology	Specific issues addressed by laws and regulations
		CLONING AND ANIMAL WELFARE • *Ethical Aspects of Animal Cloning for the Food Supply*. Opinion No. 23, January 16, 2008. European Group on Ethics in Science and New Technologies (EGE). GENERAL WELFARE OF RESEARCH ANIMALS • Directive 86/609/EEC. • Directive 2010/63/EU. GENERAL WELFARE OF FARMING ANIMALS • Directive 98/58/EC.
MIDDLE EAST		
Israel	No laws and regulations specific to GM animal production or cloning of animals; general animal welfare laws cover animals in research	GENERAL WELFARE OF RESEARCH ANIMALS • Animal Welfare Act of 1994 (covers experiments in research animals and animals generally).

(*Continued*)

Statutes and regulations in selected countries and geopolitical units affecting animals produced by modern biotechnology

Country/ geopolitical unit	Laws/regulations specifically addressing animals produced by modern biotechnology	Specific issues addressed by laws and regulations
NORTH AMERICA		
Canada	Laws and regulations specific to production of GE animals and animal clones; specific to GE animal welfare	GE ANIMAL PRODUCTION • 1993 Federal Regulatory Framework for Biotechnology established that, rather than creating new regulations, novel products produced through biotechnology will be regulated under existing regulations that cover traditional products. • Canadian Food & Drugs Act and Regulations (1999), Division 28: Novel Foods. • Health Canada is mandated to assess the safety of foods for human consumption, including genetically modified organisms (GMOs) in foodstuff, and for authorizing them to be sold in Canada. • Fish products of biotechnology are assessed before marketing by Environment Canada and Health Canada (New Substance Notifications for fish products of biotechnology) under authority of the Canadian Environmental Protection Act (1999). ANIMAL CLONE PRODUCTION • 2003 Food Directorate Interim Policy on Foods from Cloned Animals identifies food from cloned animals as novel foods. • Health Canada has yet to institute guidance although it has an interim policy document (https://www.canada.ca/content/dam/hc-sc/migration/hc-sc/fn-an/alt_formats/hpfb-dgpsa/pdf/legislation/pol-cloned_animal-clones_animaux-eng.pdf).

(*Continued*)

Statutes and regulations in selected countries and geopolitical units affecting animals produced by modern biotechnology

Country/ geopolitical unit	Laws/regulations specifically addressing animals produced by modern biotechnology	Specific issues addressed by laws and regulations
		GE ANIMAL WELFARE • CCAC guideline 1997 (updated in 2012). GENERAL RESEARCH ANIMAL WELFARE (FEDERAL) • Sections 444 to 447 of the Criminal Code of Canada protect animals (in general) from cruelty, abuse, and neglect. • Canadian Council on Animal Care (CCAC) established in 1968; certification is a condition for all institutions that receive funding from the federal granting agencies (Canadian Institutes for Health Research [CIHR] and the Natural Sciences and Engineering Research Council [NSERC]) for animal-based projects. GENERAL RESEARCH ANIMAL WELFARE (PROVINCES) • Although all provinces have laws concerning animal welfare, only certain provinces have legislation which specifically addresses animals acquired and used for scientific purposes. • Ontario has the only stand-alone law (Animals for Research Law 1990). • Many provinces do point to the CCAC with respect to welfare protection for research animals. • Saskatchewan passed the Animal Protection Act in 2018 that provides protection to all animals, including research animals. • In 2021, Quebec passed the Animal Welfare and Safety Act that applies to all animals, including those for research and states "animals are sentient beings that have biological needs."

(Continued)

Statutes and regulations in selected countries and geopolitical units affecting animals produced by modern biotechnology

Country/ geopolitical unit	Laws/regulations specifically addressing animals produced by modern biotechnology	Specific issues addressed by laws and regulations
United States	Policy specific to production of GE animals and clones; but animal welfare laws not specific to status of genetic alteration	GE ANIMAL PRODUCTION • The Office of Science and Technology Policy (OSTP) published *the U.S. Coordinated Framework for the Regulation of Biotechnology* in 1986, setting forth federal regulatory policy for ensuring the safety of biotechnology products while not impeding technology. It was updated in 1992 and 2017 to reflect changes in technology. US approach is risk-based that focuses on the characteristics of the product and the environment into which it is being introduced, not the process by which the product is created (product over process). • The FDA Center for Veterinary Medicine (CVM) *"Draft Guidance for Industry: Regulation of Intentionally Altered Genomic DNA in Animals" (#187)* outlines the role of different FDA centres in GE animal development and use (*i.e.* CDER, CBER, CFSAN, *etc.*). • FDA's Center for Biological Evaluation and Research published *"Points to Consider in the Manufacture and Testing of Therapeutic Products for Human Use Derived from Transgenic Animals (1995)"* and *"Source Animal, Product, Preclinical, and Clinical Issues Concerning the Use of Xenotransplantation Products in Humans (2003)"*. • Guidance from FDA's CDRH applies to medical devices that may incorporate GE animal material *"Medical Devices Containing Materials Derived from Animal Sources (Except for In Vitro Diagnostic Devices) – Guidance for Industry and Food and Drug Administration Staff (2016)"*.

(Continued)

Statutes and regulations in selected countries and geopolitical units affecting animals produced by modern biotechnology

Country/ geopolitical unit	Laws/regulations specifically addressing animals produced by modern biotechnology	Specific issues addressed by laws and regulations
		ANIMAL CLONE PRODUCTION • January 2008 Guidance from FDA entitled *"Use of Animal Clones and Clone Progeny for Human Food and Animal Feed"*. GENERAL RESEARCH ANIMAL WELFARE • Animal Welfare Act (AWA) in 1966 [administered by USDA]. • Amendments to the AWA (1985). • Health Research Extension Act (Public Law 99–158, November 20, 1985, codified at 42 USC §289d) [administered by the US National Institutes of Health and Public Health Service/NIH-PHS]. • Good Laboratory Practice regulations (21 CFR Part 58). • Voluntary accreditation by the Association for the Assessment and Accreditation of Laboratory Animal Care International (AAALAC).
Mexico	Laws and regulations specific to production of GE animals and specific to GE animal welfare; no laws and regulation related to animal cloning	GM ANIMAL PRODUCTION • Law on Biosecurity of Genetically Modified Organisms (GMO Law) (2005). • Provides rules on research concerning, and the release, commercialization, exportation, and importation of, Genetically Modified Organisms (GMOs). • Goal of the law is to prevent, avoid, or reduce the risks that these activities may cause to human health, the environment, biological diversity, or the health of plants and animals.

(Continued)

Statutes and regulations in selected countries and geopolitical units affecting animals produced by modern biotechnology

Country/ geopolitical unit	Laws/regulations specifically addressing animals produced by modern biotechnology	Specific issues addressed by laws and regulations
		GM ANIMAL WELFARE • Mexico's Law on Biosecurity of Genetically Modified Organisms (GMO Law) (2005). • Law addresses animal health generally as part of the overall assessment by regulatory authorities. GENERAL RESEARCH ANIMAL WELFARE • NOM-062-ZOO-1999, Technical Specifications for the Production, Care and Use of Laboratory Animals (1999). • Act of General Health and Its Regulation (2015). • Act of Animal Welfare (2002). • Mexican Official Norm for Disease Prevention and Control (2007). • Health Specifications for Canine Control Centers (2006).
SOUTH AMERICA		
Argentina	Laws and regulations specific to production of GM animals; no laws or regulations related to animal cloning; animal welfare laws are not specific to method of production	GM ANIMAL PRODUCTION • Regulatory process was established for GM plants and animals overseen by the Ministry of Agriculture, Livestock and Fisheries (1991), Directorate of Biotechnology (DB), the national Advisory Commission on Agriculture Biotechnology (CONABIA), and the National Service of Agrifood Health and Quality (SENASA). • Resolution 79-E/2017, published in the Official Gazette on November 2017 updated existing law to address GM animals and cloning. • Argentina produces both GM animals and cloned animals.

(*Continued*)

Statutes and regulations in selected countries and geopolitical units affecting animals produced by modern biotechnology

Country/ geopolitical unit	Laws/regulations specifically addressing animals produced by modern biotechnology	Specific issues addressed by laws and regulations
		GENERAL RESEARCH ANIMAL WELFARE • Animal Protection Law 14346 (1954) addresses animals of all kinds including research animals. • Law 22.421 (1981) specifically addresses welfare of wild animals.
Brazil	Laws and regulations for GM plants applied; no specific laws related to animal cloning; animal welfare laws are not specific	GE ANIMAL PRODUCTION • Regulated like plant products under Law #11,105 (March 25, 2005) as modified by two additional laws (Law #11,460 of 2007 and Decree #5,591 of 2006) by the National Technical Biosafety Commission (CTNBio) which is the body involved. GENERAL RESEARCH ANIMAL WELFARE • Specific protection of research animals from Law 11.794 (issued in 2008 and enacted by Federal Decree 6899 in 2009).
Chile	No laws and regulations specific to GM animal production or cloning of animals; general animal welfare laws exist	GENERAL RESEARCH ANIMAL WELFARE • Animal Protection Law 20380 (2009): the law establishes rules regarding animal welfare, identifies animals as sentient beings.

(*Continued*)

Statutes and regulations in selected countries and geopolitical units affecting animals produced by modern biotechnology

Country/ geopolitical unit	Laws/regulations specifically addressing animals produced by modern biotechnology	Specific issues addressed by laws and regulations
FAR EAST		
Japan	Laws and regulations specific to production of GE animals; no laws or regulations related to animal cloning; animal welfare laws are not specific	GM ANIMAL PRODUCTION • Regulation of GM plants to be applied for commercialization of GM livestock animals and insects. For production or environmental release of ME animals, MAFF's "Law Concerning the Conservation and Sustainable Use of Biological Diversity through Regulations on the Use of Living Modified Organisms" will be applied. • The Food Sanitation Act [MHLW] covers food safety aspect of GM animals. GENERAL RESEARCH ANIMAL WELFARE • 1973 Act on Welfare and Management of Animals; amended in 2005 to create new basic guidelines for experimentation (research animals) based on the Three Rs (refine, replace, reduce) for animal testing, excluding fish.

(*Continued*)

Statutes and regulations in selected countries and geopolitical units affecting animals produced by modern biotechnology

Country/ geopolitical unit	Laws/regulations specifically addressing animals produced by modern biotechnology	Specific issues addressed by laws and regulations
China	Laws and regulations specific to production of GM animals; no laws or regulations related to animal cloning; animal welfare laws are not specific	GE ANIMAL PRODUCTION • In China, restrictions on genetically modified organisms (GMOs) are primarily provided by the agricultural GMO regulations enacted by the State Council in 2001 and relevant administrative rules. The agricultural GMO regulations regulate not only crops but also animals, microorganisms, and products derived from these sources. GENERAL RESEARCH ANIMAL WELFARE • Regulations for the Administration of Laboratory Animals (1988). • Guidelines for the Humane Treatment of Laboratory Animals (2006) [first government statement that mentions "animal welfare"]. • The Ministry of Science and Technology (MOST) is the oversight body for laboratory animal research. • The China Food and Drug Administration Good Laboratory Practice (GLP) regulations have requirements concerning animal facilities and animal care and use for animals employed in GLP studies (China Food and Drug Administration 2003). • The Chinese Veterinary Medicine GLP regulations (2015) mandate the establishment of an institutional animal welfare review body and basic animal husbandry requirements in nonclinical research of veterinary medicines (Ministry of Agriculture).

(*Continued*)

Statutes and regulations in selected countries and geopolitical units affecting animals produced by modern biotechnology

Country/ geopolitical unit	Laws/regulations specifically addressing animals produced by modern biotechnology	Specific issues addressed by laws and regulations
South Korea	Laws and regulations specific to production of GM animals; no laws or regulations related to animal cloning; animal welfare laws are not specific	GM ANIMAL PRODUCTION • Living Modified Organisms (LMO) Act and regulations finalized March 2006; implemented January 1, 2008, following ratification of the Cartagena Protocol on Biosafety. • LMO Act/regulations apply to GM animals; no specific regulations have been established for the management of GM animals. • Pharmaceuticals produced from GM animals governed by the Pharmaceuticals Affairs Act. GENERAL RESEARCH ANIMAL WELFARE • Animal Protection Act of 1991 [amended several times with focus on cruelty to companion animals not laboratory animals].
India	Laws and regulations specific to production of GE animals; no laws or regulations related to animal cloning; animal welfare laws are not specific	GM ANIMAL PRODUCTION • The 1986 Environment Protection Act (EPA) governs the research, development, commercial use, and imports of GM plants, animals, their products, and byproducts (see Annex 1). • Prior to approval or importation, the Genetic Engineering Appraisal Committee (GEAC) must appraise all products derived from biotech plants and animals or other biotech organisms. • EPA 1986.

(*Continued*)

Statutes and regulations in selected countries and geopolitical units affecting animals produced by modern biotechnology

Country/ geopolitical unit	Laws/regulations specifically addressing animals produced by modern biotechnology	Specific issues addressed by laws and regulations
		GENERAL RESEARCH ANIMAL WELFARE • PCA Act of 1960 (specifically addresses use of animals in research). • The Committee for the Purpose of Control and Supervision of Experiments on Animals (CPCSEA) was established through the PCA Act. • Breeding of and Experiments on Animals Rules (1998). • Indian Veterinary Council (IVC) Act of 1984 and 1992 amendments to the IVC Act.
OCEANIA		
Australia	Laws and regulations specific to production of GM animals; laws and regulations specific to cloned animal production; specific GE and cloned animal welfare laws exist as well	GE ANIMAL PRODUCTION • The Gene Technology Act (2000) and the Gene Technology Regulations (2001) relate to GM and animal clones. Oversight by the Office of Gene Technology Regulator (OGTR). • GM and clones subject to state and territory government animal welfare legislation applicable to animals used for scientific purposes, as well as the Guidelines to promote the well-being of animals used for scientific purposes, and GCM and cloned animals for scientific purposes. ANIMAL CLONE PRODUCTION • The Department of Agriculture covers animal health (biosecurity) issues in their import risk assessments (IRAs). Cloned animals or products from cloned animals are not considered to be an animal health or biosecurity risk and have not been assessed as a hazard in the IRAs.

(*Continued*)

Statutes and regulations in selected countries and geopolitical units affecting animals produced by modern biotechnology

Country/ geopolitical unit	Laws/regulations specifically addressing animals produced by modern biotechnology	Specific issues addressed by laws and regulations
		• Food from cloned animals is not regulated in the same way as food from GE animals. Regulators consider that food products from cloned animals and their offspring are as safe as food products from conventionally bred animals and do not require any additional regulation. GE ANIMAL WELFARE • The nationally accepted Code of Conduct adopted by all States and Territories (*2013 Australian Code for the Care and Use of Animals for Scientific Purposes*) has provisions for welfare of GE animals. This code is stated to apply to all live non-human vertebrates and cephalopods. ANIMAL CLONE WELFARE • Nationally Code of Conduct adopted by all States and Territories (*2013 Australian Code for the Care and Use of Animals for Scientific Purposes*); applies to all live non-human vertebrates and cephalopods. • Provisions for welfare of clones and GM animals. GENERAL RESEARCH ANIMAL WELFARE (FEDERAL) • No national laws solely for animal welfare; all states and territories regulate animal welfare under their jurisdiction.

(*Continued*)

Statutes and regulations in selected countries and geopolitical units affecting animals produced by modern biotechnology

Country/ geopolitical unit	Laws/regulations specifically addressing animals produced by modern biotechnology	Specific issues addressed by laws and regulations
		GENERAL RESEARCH ANIMAL WELFARE (STATES AND TERRITORIES) • Apart from New South Wales, each State and Territory has incorporated provisions relating to animal research in its anti-cruelty statute. • New South Wales (NSW) Prevention of Cruelty to Animals Act (1979) and the Research Animals Act (1985). • Australian Capital Territory (ACT) Animal Welfare Act (1992). • Victoria Prevention of Cruelty to Animals Act (1986). • Queensland Animal Care and Protection Act (2001). • Northern Territory Animal Welfare Act. • Western Australia Animal Welfare Act (2002). • South Australia Animal Welfare Act (1985). • Tasmania Animal Welfare Act (1993).
New Zealand	Laws and regulations specific to production of GM animals; no specific laws and regulations for cloning of animals; specific GM and clone welfare laws	GE ANIMAL PRODUCTION • The Hazardous Substances and New Organisms (HSNO) Act 1996 relates to GM plants and animals; • Enabling regulations (2003) known as the Hazardous Substances and New Organisms (Low-risk Genetic Modification) Regulations. GE ANIMAL WELFARE • Animal Welfare Amendment Act (No. 2) 2015.

(Continued)

Statutes and regulations in selected countries and geopolitical units affecting animals produced by modern biotechnology

Country/ geopolitical unit	Laws/regulations specifically addressing animals produced by modern biotechnology	Specific issues addressed by laws and regulations
		ANIMAL CLONE WELFARE • Animal Welfare Amendment Act (No. 2) 2015. GENERAL RESEARCH ANIMAL WELFARE • Animal Welfare Act of 1999. • Animal Welfare Amendment Act (No. 2) 2015 [recognizes animals as sentient beings].
Singapore	Laws and regulations specific to production of GM animals; no specific laws and regulations for animal cloning; no specific welfare laws for these animals	GM ANIMAL PRODUCTION • Approval processes for animal biotechnology and plant biotechnology are the same. • The Singapore Food Agency (SFA) replaced the Agri-Food and Veterinary Authority (AVA) as the national body that regulates GM market (2019). • Animal Veterinarian Services (AVS) manages all non-food animal matters. • Multi-agency Genetic Modification Advisory Committee (GMAC) established under the country's Ministry of Trade and Industry in 1999 to provide science-based advice on research, development, production, release, use, and handling of GE events in Singapore. SFA considers GMAC's recommendations before making an official regulatory decision. • Currently, animal biotechnology development is limited to research activities in fish hatchery technology. There is no commercial animal biotechnology production in the country.

(Continued)

Statutes and regulations in selected countries and geopolitical units affecting animals produced by modern biotechnology

Country/ geopolitical unit	Laws/regulations specifically addressing animals produced by modern biotechnology	Specific issues addressed by laws and regulations
		GENERAL RESEARCH ANIMAL WELFARE • 2004 NACLAR Guidelines were the first standards in Singapore to apply to animals in research and would include GE animals and animals produced by cloning.
Malaysia	Laws and regulations specific to production of GE animals; no specific laws and regulations for cloning of animals; no specific GM or animal clone welfare laws exist	GE ANIMAL PRODUCTION • Biosafety Act and 2010 Approval Regulations (2007). • The Islamic Development Authority of Malaysia (JAKIM) opposes animal biotechnology products for the purpose of consumption in Malaysia. GENERAL RESEARCH ANIMAL WELFARE • Animals Act of 1953. • Malaysian Fisheries Act 1985. • Wildlife Conservation Act of 2010. • Animal (Amendment) Act 2013. • Malaysian Animal Welfare Act 2015. • The Malaysian Code for the Care and Use of Animals for Scientific Purposes (MyCode).

Index

Note: **Bold** page numbers refer to tables; *italic* page numbers refer to figures and page numbers followed by "n" denote endnotes.

For Product Safety Concerns and Information please contact our EU representative GPSR@taylorandfrancis.com Taylor & Francis Verlag GmbH, Kaufingerstraße 24, 80331 München, Germany